A FIRST COURSE IN NUMBER THEORY

A FIRST COURSE IN NUMBER THEORY

Hugh M. Edgar

San Jose State University

Wadsworth Publishing Company

Belmont, California

A Division of Wadsworth, Inc.

Mathematics Publisher: *Kevin J. Howat*
Development Editor: *Anne Scanlan-Rohrer*
Assistant Editor: *Barbara Holland*
Editorial Assistant: *Sally S. Uchizono*
Production Editor: *Gary Mcdonald*
Managing Designers: *Merle Sanderson, James Chadwick*
Print Buyer: *Karen Hunt*
Designer: *Lisa S. Mirski*
Copy Editor: *Steven Gray*
Compositor: *Graphic Typesetting Service*
Cover: *Lisa S. Mirski*
Signing Representative: *Tina Allen*

Printed in the United States of America 19

1 2 3 4 5 6 7 8 9 10---92 91 90 89 88

Library of Congress Cataloging-in-Publication Data

Edgar, Hugh M., 1934–
A first course in number theory / Hugh M. Edgar.
p. cm.
Bibliography: p.
Includes index.
ISBN 0-534-08514-8
1. Numbers, Theory of. I. Title
QA241.E34 1988
512′.7--dc19 87-17756
CIP

To Leo Moser, *maître et ami*

Contents

Preface

My love of number theory goes back to Leo Moser at U. at A. (Alberta) in the 1956–57 academic year, when I was half of Leo's Math 110 class—a first-semester graduate-level course in number theory. Jim Riddell, now at U. Vic (Victoria, B.C.), was the other half of the class. The two other mathematicians who influenced me most as a student were Sarvadaman Chowla, my Ph.D. thesis advisor at CU (University of Colorado), and Louis Joel Mordell, from whom I took algebraic number theory at CU in the 1959–60 academic year. I am keenly aware of my great debt to these three mathematicians.

The objectives of this book can be roughly broken down into eight main items. First, I want to present—as clearly as possible—what I consider to be the core material for a one-semester first course in number theory, typically at the junior/senior level in a college or university. I have deliberately not succumbed to the considerable lure of, for instance, the currently popular topics of public-key cryptography and primality testing in the interest of writing a truly one-semester text. For the same reason I haven't systematically developed the extremely appealing subject of continued fractions, but, rather, have introduced this topic via problems.

Second, I have made an overt effort to keep the style of the book informal, without compromising the integrity of the mathematics. I want to try to communicate to the greatest possible extent with the student/reader. It is a well-known fact that mathematics is not a spectator sport, so (for instance) not all the loose ends of a proof will, in general, be cleaned up by me; rather, the reader is encouraged strongly to take an active role in making the proof complete. Some good models for this type of activity are presented early in the text.

Third, the problem sets at the end of each section are the heart of the book. I have tried to include enough problems of various types: routine applications of the material developed in the body of the text; true/false questions (designed to run a check on how carefully the material has been read and—more adventurously—to stimulate the student to transcend the text); little proofs; and, more rarely, some genuinely difficult problems. These are identified by a dagger (†). One strategy I have used in creating the problem sets is illustrated by problem 3(d) on page 68 in conjunction with problem 22 on page 70: I try to get the student to formulate a natural conjecture (or guess at a fact) early in the problem set, and then I reinforce the correct answer to the true/false question by stating the correct answer in a later problem and asking for a proof. Those problems for which I have provided hints or a solution in the Solutions/Hints to Selected Problems section are indicated by an asterisk (*).

Fourth, it is extremely important to dispel the "bag of tricks" myth which tends to be leveled at the subject of elementary number theory. I have tried to emphasize the many recurring themes: use of the Division Algorithm as a proof tool; integral linear combinations; applications of the principle of the least element; pairing processes; etc.

Fifth, an important feature of this text that sets it apart from all others (to the best of my knowledge) is the early introduction and subsequent reliance upon p-adic valuations. It seems to me that every 1988 student of number theory should be exposed to Kurt Hensel's wonderful creation in the first course. The case for local methods is abundantly documented in the higher reaches of mathematics (not, mind you, in number theory only!), and it seems imperative that at least the rudiments be included at this level. This approach is perhaps best illustrated in Chapter 4, Section 4.2 where the p-adic version of Newton's method of successive approximations is presented. The inclusion of $Z[\sqrt{-10}]$ as an example of a context where there is no Fundamental Theorem of Arithmetic is not by any means novel, but I believe that it is an extremely important thing to do. Failure to do so, in my opinion, is an arithmetical sin analogous to failing to inform the student of the existence of non-Euclidean geometries in a geometry course.

Sixth, I have made a strenuous attempt to have the subject material of the text flow smoothly from one section to the next and, more globally, from each chapter to its immediate successor. The most important thing I did along these lines was to choose the contents of each section with great care. The other devices used are: introductory remarks in individual sections and at the beginning of each chapter; and numerous back-references and (less numerous) forward references, involving both material in the body of the text and problems.

Seventh, I want very much to advertise the fact that mathematics in general and number theory in particular is an exciting, ongoing activity. I have used some of the notes at the ends of the chapters to acquaint the reader with some of the spectacular recent number-theoretic advances and breakthroughs: the proof of Mordell's Conjecture by Gerd Faltings; various significant advances in connection with "Fermat's Last Theorem"; near resolution of Artin's Conjecture; etc. These examples of research in number theory are at a very advanced level and thus are totally inappropriate in the body of a text such as this. However, I hope that some students will be stimulated to increase their mathematical horizons significantly through additional reading. Instructors can use these notes to extend or enhance the course as they see fit.

Eighth, the two appendixes. Appendix I – Basics is designed to make the text as self-contained as possible. It includes material on equivalence relations, (mathematical) induction and binomial coefficients, as well as comments on the idea of a counterexample and the technique of proof by contradiction. Appendix II – Abstract Algebra is designed to briefly introduce the reader to (or reacquaint the reader with) the world of abstract algebra and to point out that elementary number theory and elementary abstract algebra are intimately related subjects.

Suggestions for the Use of This Book: My main suggestion is that the student/reader study the book "with pencil in hand." The way to learn this material is by doing it, which in this case means doing as many problems as possible and being a full participant in making sure that there is an airtight proof for each theorem in the text. I also think it is important to *read* all of the problems. It isn't really reasonable to expect to do all of the problems, but you can pick up a lot of information quite quickly by scanning a problem set, and (with any luck at all) a certain amount of osmosis will take place.

Acknowledgments. A large number of people have given me help in various ways: Joe Buhler, Tatiana Deretsky, Don Duncan, Don Fiehmann, Basil Gordon, Paul Halmos, Peter Hilton, Brad Jackson, Jean Pedersen, Marilyn Ruch, Jim Smart, Dmitri Thoro, and my number theory students from the Fall 1985 semester (who were subjected to the first draft), together with my number theory students from the Spring 1986 semester (who probably suffered some degree of schizophrenia induced by having a combination of the first and second drafts).

Dan Goldston, my officemate, deserves special mention: he taught me everything I know about MacWrite (which is still a minuscule amount) and generously allowed me complete access to his Macintosh system; he also made me back up disks in case I crashed!

I would also like to thank the various people at Wadsworth who were involved in the making of this book: Jim Harrison, Anne Scanlon-Rohrer, Kevin Howat, and Tina Allen. I particularly want to thank Steven Gray for his meticulous job of copyediting—I think the book is the better for it. Finally, I want to thank the Production Editor of my text, Gary Mcdonald. His expertise and efficiency were of great value to me, particularly in the "terminal" stages.

By far my greatest debt is to my wife, Kathleen. She encouraged me to write this text, she typed the first draft, and has been most supportive throughout the project. Her artistic suggestions contributed greatly to the cover design.

A FIRST COURSE IN NUMBER THEORY

1 Divisibility Without Primes

1.1 Divisibility and the Division and Euclidean Algorithms

The first concept we introduce is that of divisibility.

Definition 1.1.1: An integer c is said to be **divisible** by an integer b if and only if there is exactly one integer—call it x—such that $c = bx$.

If c is divisible by b we write $b|c$. The notation $b \nmid c$ means that c *isn't* divisible by b. Among the many other expressions that mean the same thing as "c is divisible by b" are: "b divides c," "b is a divisor of c," "c is a multiple of b," and "b is a factor of c."

The definition requires that there be precisely one value of x for which $c = bx$. This is important. For instance, suppose we think that $0|c$ for some integer c. Then the equation $c = bx = 0x$ means that c would have to be zero, but now we are in trouble because $0 = 0x$ doesn't have exactly one solution for x. The conclusion is that we cannot ever use $b = 0$ in the definition of divisibility; i.e., it doesn't make sense to divide by zero. So $16|64$, $0 \nmid c$, regardless of the choice of the integer c and $105 \nmid 24$.

The notion of divisibility will be extremely important throughout this book, so we immediately begin to establish some of the basic properties of this concept. It is also important to know some of the "non-properties" of the divisibility relation—things that can't be claimed as properties of the divisibility relation.

We make a start by collecting several divisibility properties in the following little theorem.

Theorem 1.1.1:

1. $b|c$ if and only if $\pm b|\pm c$.
2. If $b|c$ and $c|d$, then $b|d$ (the **transitive** property of divisibility).
3. If $b|c$ and c is positive, then $b \leqslant c$.
4. If $b|c$ and $b|d$, then $b|(cx + dy)$, regardless of the choice of the integers x and y (the **integral linear combination** property of divisibility).

Proof: Since this is the first example of a proof in this text, I would like to do a thorough job of what I do. So I will do a thorough job of proving parts 2 and 4 of the theorem, leaving the job of providing thorough proofs of parts 1 and 3 to you (see page 6, problem 11).

The conclusion we are after in part 2 of the theorem is $b|d$; i.e., appealing immediately to the one and only tool available to us at present—the definition of *divisibility*—we need to prove that there exists a uniquely determined integer, say z, such that $d = bz$. Now that we know what we're looking for, we may have a better idea of how to go about finding it. Specifically, since we discovered our need for z by converting the desired conclusion $b|d$ into the corresponding equation $d = bz$, it is reasonable to do likewise with our two assumptions. The assumption $b|c$ becomes $c = bx$, where x is a uniquely determined integer. Likewise $c|d$ becomes $d = cy$ for some uniquely determined integer y. If we now replace c by bx in the equation $d = cy$, we get $d = (bx)y = b(xy) = bz$ with $z = xy$: z is a uniquely determined integer because each of x and y was. The proof of part 2 is complete.

In part 4 we are given $c = bu$, $d = bv$ for unique integers u and v. Hence $cx + dy = (bu)x + (bv)y = b(ux + vy)$; and once x and y are given, $ux + vy$ is, again, a uniquely determined integer, since u and v are unique integers.

Remarks: **1.** I deliberately included every step in the proof of part 2 of Theorem 1.1.1 because it was our first example of a proof in this text. The proof of part 4 was made much shorter because the proof technique is identical to that employed in part 2. **2.** Part 1 of Theorem 1.1.1 tells us that divisibility is insensitive to algebraic sign. A numerical example will make this very clear. Since $(-18)|(+90)$, it follows from part 1 that $(-18)|(-90)$, $(+18)|(-90)$, and $(+18)|(+90)$. Usually it is convenient to work with positive integers, in which case we would just say $18|90$. The upshot of this is that, when we are working with divisibility, it often suffices to confine our attention to the positive integers. **3.** It is virtually impossible to overemphasize the importance and usefulness of part 4 of Theorem

1.1.1, the result on integral linear combinations. In spite of being very easy to prove, it serves as a powerful theoretical weapon—a fact that we shall illustrate over and over again.

We give over the rest of this section to an initial discussion of two intimately related algorithms: the Division Algorithm, and the Euclidean Algorithm. The Division Algorithm is a very homey and familiar fact. Schematically, we have

$$c\overline{)b}^{\;q}{}_{r}$$

which can be interpreted as $b/c = q + r/c$. A precise formulation is given in the following theorem.

Theorem 1.1.2 – The Division Algorithm: If b and c are positive integers, then there exist uniquely determined integers q (the quotient) and r (the remainder) such that $b = cq + r$ with $0 \leqslant r < c$.

Proof: Let S denote the set of all integers of the form $b + cz$, where z runs through the entire set of integers. S contains nonnegative integers—for example, $b + c$. Hence, S must possess its uniquely determined least nonnegative element.[1] Let's denote this least nonnegative element of the set S by the symbol r. We may also choose our notation so that this special element r is obtained by putting $z = -q$, so that $b + c(-q) = r$ or $b = cq + r$. Now $r - c = (b + c(-q)) - c = b + c(-q - 1)$ is an element of S that is less than r and hence necessarily negative. Hence $r < c$, as stated in the theorem. To complete the proof, we need to show that, if $b = cq' + r'$ holds with $0 \leqslant r' < c$, then necessarily $q = q'$ and $r = r'$. Suppose, for instance, that $q' < q$. Then $r' = b + c(-q') \geqslant b + c(-q + 1) = (b + c(-q)) + c = r + c \geqslant c$ so that $r' \geqslant c$. But this is nonsense because part of the assumption about the pair (q', r') was that $0 \leqslant r' < c$. So $q' < q$ is impossible and so we must have $q' \geqslant q$. If we interchange the roles of q' and q in the argument, however, we can likewise conclude that $q \leqslant q'$, and so $q = q'$ is forced. Finally, $q = q'$ implies that $r = r'$, and so the uniqueness is proved.

The Euclidean Algorithm is a clever way of repeatedly using (the standard mathematical term is **iterating**) the Division Algorithm. As we shall prove in Section 1.2, this technique provides an efficient way to find the so-called GCD (greatest common divisor) of the two input positive integers b and c. We first illustrate the Euclidean Algorithm for a specific choice of b and c, and we then write out the general notational setup for use in the problems at the end of this

[1]Here we are using the so-called Principle of the Least Element; see Appendix I, page 106.

section (see, specifically, page 5, problems 2 and 3) and in the theoretical results of the next section.

Example 1.1.1: If $b = 741$ and $c = 715$, then one application of the Division Algorithm gives $b = cq + r$, with $0 \leqslant r < c$, which in this case becomes $741 = (715 \cdot 1) + 26$ with $0 \leqslant 26 < 715$. We now make a second application of the Division Algorithm, with $715 = c$ playing the role of b and $26 = r$ playing the role of c. So $715 = (26 \cdot 27) + 13$, with $0 \leqslant 13 < 26$. Shifting left once more—letting 26 play the role of b and letting 13 play the role of c—gives $26 = (13 \cdot 2) + 0$. The appearance of zero signals the end of the algorithm. For one thing, if we tried to continue, we would be trying to solve $13 = (0 \cdot (\ )) + \{\ \}$, with $0 \leqslant \{\ \} < 0$, which is impossible. The Euclidean Algorithm for this example can conveniently be written as

$$\begin{aligned} 741 &= (715 \cdot 1) + 26 \\ 715 &= (26 \cdot 27) + 13 \\ 26 &= (13 \cdot 2) + 0 \end{aligned}$$

As we shall see in the next section, this algorithm will tell us that the greatest common divisor (GCD) of the two positive integers 741 and 715 is 13. Already, from this display, we can show that 13 can be written as an integral linear combination of the two input positive integers 741 and 715: $13 = 715 - (26 \cdot 27) = 715 - (741 - 715) \cdot (27) = (-27) \cdot (741) + (+28) \cdot (715)$.

Here is the general notational setup:

$$\begin{aligned} b &= cq_1 + r_1, & 0 < r_1 < c \\ c &= r_1q_2 + r_2, & 0 < r_2 < r_1 \\ r_1 &= r_2q_3 + r_3, & 0 < r_3 < r_2 \\ &\ \vdots \\ r_{k-2} &= r_{k-1}q_k + r_k, & 0 < r_k < r_{k-1} \\ r_{k-1} &= r_kq_{k+1} + 0 \end{aligned}$$

Remark: If r_k, for instance, were zero, then the algorithm would have terminated earlier. The notation is arranged to advertise r_k as the last (and least) positive remainder.

PROBLEMS

Solutions (or hints at solutions) to the problems marked with an asterisk (*) are given on page 121.

1. Suppose we have three integers b, c, and d satisfying $b \nmid c$ and $c \nmid d$. Must it be the case that $b \nmid d$? If your answer is no, justify your answer by pro-

ducing a numerical **counterexample**[1]—i.e., three integers b, c, and d for which $b \nmid c$, $c \nmid d$, and yet $b|d$. If your answer is yes, justify your answer by giving a proof.

***2.** Apply the Euclidean Algorithm to the following pairs of positive integers (as stated previously, this process will produce the GCD (greatest common divisor) of the input pair of positive integers):
(a) $b = 288$, $c = 51$.
(b) $b = 1{,}687$, $c = 482$.
***(c)** $b = 3{,}587$, $c = 1{,}819$.
(d) $b = 144$, $c = 89$.

***3.** Express G as an integral linear combination of b and c for the following triples of positive integers:
(a) $G = 3$, $b = 288$, $c = 51$.
(b) $G = 241$, $b = 1{,}687$, $c = 482$.
***(c)** $G = 17$, $b = 3{,}587$, $c = 1{,}819$.
(d) $G = 1$, $b = 144$, $c = 89$.

[HINT (strong!): Use the work you have done in problem 2, and mimic Example 1.1.1 on page 4.]

4. Men-in-Jail Problem: There are N men in cells numbered 1, 2, 3, . . . , N in a linear cell block:

1 2 3 4 5 . . . N

The warden and the inmates agree to perform the following N-day experiment. On the first day of the experiment, the warden unlocks everybody's cell door. On the second day, the warden locks the cell door of every second inmate (i.e., numbers 2, 4, 6, 8, . . .). On the third day, the warden turns the key in the cell door of every third inmate (i.e., numbers 3, 6, 9, . . .). For instance, the man in cell number 3 has his cell locked on the third day, whereas the man in cell number 6 has his cell unlocked on the third day. In general, on the kth day, for every value of k such that $1 \leqslant k \leqslant N$, the warden turns the key in every kth cell (i.e., numbers k, $2k$, $3k$, . . .).

The gentlemen's agreement is: (a) Nobody will attempt to escape during the course of the experiment, and (b) an inmate gets out of jail at the end of the N-day experiment if and only if his cell door is unlocked at that time. Your problem is to describe precisely who goes free at the end of the N-day experiment. [HINTS: (1) Determine the fate of, say, the first ten inmates. This process will probably already stimulate you to guess the correct answer. (2) $1 \leftrightarrow 24$, $2 \leftrightarrow 12$, $3 \leftrightarrow 8$, $4 \leftrightarrow 6$.[2]]

[1]See Appendix I, page 107.

[2]The notation $b \leftrightarrow c$ means "b corresponds to c."

*5. In each of the following, decide whether the statement is true or false. If you think the statement is true, justify your answer by giving a proof. If the statement is false, give a numerical counterexample.

*(a) If $bc|bd$, then $c|d$.

*(b) If $b|(c^2 + 1)$, then $b|(c^4 + 1)$.

*(c) If $b|(c^2 - 1)$, then $b|(c^4 - 1)$.

*(d) If $b^2|c^3$, then $b|c$.

*(e) If $b|c$ and $d|e$, then $bd|ce$.

*(f) If $b^3|c^2$, then $b|c$.

*6. An integer b is said to be **even** if and only if $2|b$; i.e., this is the divisibility way of expressing "even-ness". Likewise, b is **odd** if and only if $2 \nmid b$ (hence, if and only if b fails to be even). Give a proof for each of the following facts.

(a) The product of any two consecutive integers is even.

(b) Every odd integer b is of the form $b = 2k + 1$ for a suitable uniquely determined integer k.

*(c) If b is odd, then $8|(b^2 - 1)$.

*7. Prove each of the following:

(a) If an integer b divides both of the integers x and y, then b divides $x + y$, $x - y$, and $y - x$.

*(b) If b and c are both positive integers and $b|c$ and $c|b$, then $b = c$.

*8. (This problem assumes some knowledge of **infinite series.**[1]) Prove that the number e is irrational. [HINTS: (1) A number q is said (defined) to be **rational** if and only if there exist integers b and c, $c \neq 0$, such that $q = b/c$. **Irrationality** is lack of rationality. (2) Use the Maclaurin expansion of the exponential function and, specifically, estimates of partial sums thereof.]

9. We define the **Fibonacci sequence** $\{F_n\}_{n=0}^{+\infty}$ as follows: $F_0 = 1$, $F_1 = 1$, and $F_n = F_{n-1} + F_{n-2}$ for all integers $n \geqslant 2$. Prove that no two consecutive Fibonacci numbers are divisible by the same integer d, regardless of the choice of $d > 1$.

10.(a) Describe precisely the integers b for which $b|b$.

*(b) Describe precisely the pairs of integers b and c for which the following is true: if $b|c$, then $c|b$.

*(c) Is the relation $\nmid$ transitive?

11. Prove parts 1 and 3 of Theorem 1.1.1 (page 2).

12. **(Finite Simple Continued Fractions.)** The Euclidean Algorithm can be reformulated in a way that pays great dividends later. For example, the

[1]See page 108, Appendix I.

arithmetic connected with Example 1.1.1 on page 4 can be advertised as follows:

$$741/715 = 1 + (26/715), \qquad 715/26 = 27 + (13/26), \qquad 26/13 = 2$$

or, in final form,

$$741/715 = 1 + 1/(27 + (1/2)).$$

Mimic this numerical example for the following ordered pairs (b, c) of positive integers:

(a) $b = 288$, $c = 51$.
(b) $b = 1{,}687$, $c = 482$.
(c) $b = 3{,}587$, $c = 1{,}819$.
(d) $b = 144$, $c = 89$.

***13.*(a)** Assume that $b > 1$ is an integer and that $m|n$, where m and n are positive integers. Must it be the case that $(b^m - 1)|(b^n - 1)$? Prove that your answer is correct.

(b) If $b > 1$ is an integer and $(b^m - 1)|(b^n - 1)$, does it follow that $m|n$? Prove that your answer is correct.

14. (The Greatest Integer Function, [ ].) If x is any real number, let us define $[x]$ to be the integer satisfying $x - 1 < [x] \leqslant x$. Thus $[5] = 5$, $[\pi] = 3$, and $[-7/2] = -4$.

(a) Prove that the quotient q in the statement of the Division Algorithm (Theorem 1.1.2 on page 3) can be written as $q = [b/c]$.
(b) State a formula for the number of inmates whose cell door locks are either locked or unlocked on the kth day in the Men-in-Jail Problem (problem 4 on page 5).

1.2 Greatest Common Divisor (GCD) and Least Common Multiple (LCM)

In the latter part of Section 1.1, I claimed, without proof, that the Euclidean Algorithm provides us with a good way of determining the greatest common divisor (GCD) of any pair of positive integers. I would now like to substantiate that claim, and doing so requires that we put the notion of greatest common divisor on a firm basis.

An integer b is said to be a **common divisor** of the two integers c and d precisely when both of the conditions $b|c$ and $b|d$ are satisfied. This preliminary definition sets us up for the most important definition of this section.

Definition 1.2.1: The positive integer G is said to be the **greatest common divisor (GCD)** of the two positive integers b and c, provided that (a) $G|b$ and $G|c$ (i.e., G is a common divisor of b and c), and (b) if $d|b$ and $d|c$, then $d|G$.

The standard notation is $G = (b, c)$, which should not (please!) be confused with the usual ordered pair notation.

Our main job is to establish the existence of the GCD; i.e., we have to prove that, regardless of the choice of the positive integers b and c, there exists at least one positive integer G satisfying conditions (a) and (b) of the definition. Notice that there can't be two such positive integers because, if G and G' both satisfy Definition 1.2.1, then $G|G'$ and $G'|G$ and hence $G = G'$ (please fill in the details missing in this argument—see problem 7(b) on page 6). Thus the uniqueness aspect of the GCD is no problem. On to our main job: suppose that b and c are arbitrary but fixed positive integers. We define S to be the set of all integral linear combinations of the input pair b and c. So a generic element of S has the form $bx + cy$, where x and y are integers. The set S contains positive integers (why?) and so must contain a least positive integer which we can denote by the symbol G. Now we must show that this positive integer G satisfies conditions (a) and (b) of the above definition and hence is the GCD of b and c.

In showing now that $G|b$ we get our first (but by no means our last!) chance to use the Division Algorithm as a proof tool: according to that theorem, we may write $b = Gq + r$ for uniquely determined integers q and r, with $0 \leqslant r < G$. If $r = 0$, then $b = Gq$ and hence $G|b$ as desired, so we only have to prove that $r > 0$ is impossible. If $r > 0$ then $0 < r = b - Gq < G$ which tells us that r is a positive integer smaller than G. However G was, among other things, an element of S, so there exist integers x_0 and y_0 such that $G = bx_0 + cy_0$. Now $r = b - Gq = b - (bx_0 + cy_0)q = b(1 - x_0) + c(-y_0q)$, so r is an integral linear combination of b and c, and so r is an element of S.

What has happened? On the assumption that r is positive, we have shown that r is a positive element of S that is less than G. But G started life as the least positive element of S, so we have a contradiction that proves the assumption $r > 0$ to be untenable. Hence $r = 0$, and $G|b$ has been proved. The argument for $G|c$ is identical to what we just went through, so we have now proved that G satisfies condition (a) of Definition 1.2.1. It is easy to prove that G satisfies condition (b):[1] if $d|b$ and $d|c$, then d divides every integral linear combination of b and c (why?); but G is an element of S and hence is an integer of this type.

Remarks: **1.** The above argument establishes both the existence and the uniqueness of the GCD, (b,c), of any pair b and c of positive integers. The principle

[1]I believe that it is extremely important to realize that some proofs are easy. The proof that G satisfies condition (b) is a proof of this fact.

of the least element is an existence and uniqueness statement. **2.** It is easy to extend the definition of GCD so as to include negative integers. **3.** We already have an extremely useful **characterization of the GCD** of any pair of integers: $G = (b, c)$ is the least positive integer that is representable as an integral linear combination of b and c.

Example 1.2.1: $(24, 60) = 12$, $(5, 6) = 1$, $(35, 105) = 35$. More generally, $(n, n + 1) = 1$, regardless of the choice of the integer n. This fact can be seen very easily by using the characterization mentioned in Remark 3: $1 = (n + 1) - n = (n + 1)x + n(y)$ with $x = +1$ and $y = -1$. Thus the number 1 is representable as an integral linear combination of $(n + 1)$ and n, and 1 is certainly the least positive integer that can be so represented because 1 is the least positive integer, period! Again it is easy to generalize from $(35, 105) = 35$ to $(b, c) = |b|$ if $b|c$, which is valid for every nonzero integer b.

It is high time to keep the promise made near the end of Section 1.1—namely, that the Euclidean Algorithm provides us with one technique (we shall have others later) for finding the GCD of two (positive) integers. For convenience, we reproduce here the general notational setup of the Euclidean Algorithm:

$$\begin{array}{lll} b & = cq_1 + r_1, & 0 < r_1 < c \\ c & = r_1q_2 + r_2, & 0 < r_2 < r_1 \\ r_1 & = r_2q_3 + r_3, & 0 < r_3 < r_2 \\ & \vdots & \\ r_{k-2} & = r_{k-1}q_k + r_k, & 0 < r_k < r_{k-1} \\ r_{k-1} & = r_kq_{k+1} + 0 & \end{array}$$

We now prove that r_k is the GCD of b and c. In order to show that r_k satisfies condition (a) of the definition of GCD, we start at the bottom of the above display and work our way up to the top; the bottom equation tells us that $r_k|r_{k-1}$. Moving up to the second equation from the bottom, we can see that $r_k|r_{k-2}$ because we already know that $r_k|r_k$ and $r_k|r_{k-1}$, and the second equation from the bottom advertises r_{k-2} as, guess what, an integral linear combination, of r_k and r_{k-1}! By now the pattern is already established. The next equation up expresses r_{k-3} as an integral linear combination of r_{k-2} and r_{k-1}, hence $r_k|r_{k-3}$. By the time we reach the second equation from the top of the display, we know (from below) that $r_k|r_2$ and $r_k|r_1$ and hence can conclude that $r_k|c$. Finally, from the top equation we learn that $r_k|b$ and hence that r_k is a common divisor of b and c; i.e., we have established that r_k satisfies condition (a) of the definition of GCD.

The proof of the fact that r_k satisfies condition (b) of the definition of GCD goes exactly the same way as the above argument—the only difference being that we start at the top of the display and work our way down. We leave the details as an exercise for that elusive creature, the interested reader.

Now that we know that every pair of positive integers has one and only one GCD and that, furthermore, we can use the Euclidean Algorithm to find (b, c), we want to discover a number of properties of the GCD that will help us with computations. Here is a list of some elementary but useful properties of the GCD. I will prove a couple of them, to illustrate some of the proof techniques involved, and then leave the proofs of the remaining properties as problems.

List of Properties of the GCD

1. $(nb, nc) = n(b, c)$ for all positive integers n, b, and c.
2. If the positive integer d is a common divisor of the positive integers b and c, then $(b/d, c/d) = (b, c)/d$.
3. $(b/(b, c), c/(b, c)) = (b, c)/(b, c) = 1$ for all positive integers b and c.
4. If $(b, c) = 1$, then $(d, bc) = (d, b)(d, c)$ for all positive integers d.
5. If $(b, d) = (c, d) = 1$, then $(bc, d) = 1$.
6. If $bx + cy = m$ for integers x and y, then $(b, c)|m$.

Proof of Property 1: By Remark 3 on page 9, (nb, nc) is the least positive integer which is representable as an integral linear combination of nb and nc, which is surely n times (the least positive integer which is representable as an integral linear combination of b and c).

Proof of Property 6: Since m is an integral linear combination of b and c and since $(b, c)|b$ and $(b, c)|c$, we get the desired conclusion by making an application of part 4 of Theorem 1.1.1 (page 2).

We say that b and c are **relatively prime** if their greatcst common divisor is 1. This situation—where two integers are relatively prime—is a very special case of the general GCD setup, but it is a very important special case that comes up a lot. One rather loose way of describing the situation $(b, c) = 1$ is to say that the two integers b and c "have nothing to do with each other" from the point of view of divisibility; i.e., the integers b and c have no divisors in common except for the numbers 1 and -1, which are divisors of every integer.

An extremely useful result that involves the concept of relative primeness is the following little theorem.

Theorem 1.2.1: If $b|cd$ and $(b, c) = 1$, then $b|d$.

Proof: The assumption $(b, c) = 1$ guarantees the existence of integers x and y for which $bx + cy = 1$. The integer d must be introduced at some point in this argument if we are to get the desired conclusion—right? And now that we have the equation $bx + cy = 1$, what could be a more natural way of getting d into the act than to multiply both sides of this equation by d? So now $(bx + cy)d = 1 \cdot d = d$. We also expect to use all the hypotheses of the theorem in making the proof, and the hypothesis $b|cd$ hasn't been used. This motivates us to rewrite the left-hand side of the previous equation so as to obtain $b(xd) + (cd)y = d$. But now $b|d$ can be deduced because $b|b(xd)$ and $b|(cd)y$, so $b|b(xd) + (cd)y = d$.

We finish this section by introducing the notion of least common multiple (LCM). Since the GCD has been worked over fairly carefully and since the two concepts are very much alike in some sense,[1] we will be briefer and less rigorous in talking about the LCM. The best way to achieve good understanding of the LCM is to do the appropriate problems at the end of this section.

An integer d is said to be a **common multiple** of the two integers b and c if both $b|d$ and $c|d$. The positive integer L is said to be the **least common multiple (LCM)** of the two integers b and c, provided that

a. L is a common multiple of b and c, and

b. If d is any common multiple of b and c, then $L|d$

The most common notation for the LCM is $L = [b, c]$. As in the case of the GCD, we are obligated to prove the existence and uniqueness of the LCM, but—having done the proof for the GCD—I am going to avoid this by making it a problem.

Here are a couple of useful properties of the LCM.

1. The set of all common multiples of b and c coincides with the set of all integral multiples of $[b, c]$.
2. $n[b, c] = [nb, nc]$ for every positive integer n.

We now prove a final result—one that provides a linkage between the two concepts of GCD and LCM.

Theorem 1.2.2: If b and c are arbitrary positive integers, then we have $(b, c)[b, c] = bc$.

Proof: The idea is to make this a two-stage process. The first stage is to argue that there is no loss of generality in assuming that b and c are relatively prime. The second stage is to show that the result is true in the case $(b, c) = 1$. My first move is to say that $(b, c)[b, c] = bc$ is true if and only if $(b, c)[b, c]/(b, c)^2 = bc/(b, c)^2$ is true; certainly it is hard to argue with that. So now we need to figure out how to establish the truth of $(b, c)[b, c]/(b, c)^2 = bc/(b, c)^2$. In order to do this, we rewrite the desired equality very carefully—namely, in

[1]See page 15, note 3.

the form $(b, c)/(b, c) \cdot [b, c]/(b, c) = b/(b, c) \cdot c/(b, c)$ which is equivalent to $[b, c]/(b, c) = b/(b, c) \cdot c/(b, c)$. (Please remember that we haven't yet established the truth of this equation; we are just trying to make our goal visible.) We now apply property 2 of the LCM, with the symbols n, b, and c replaced by (b, c), $b/(b, c)$, and $c/(b, c)$, respectively, to get $[(b, c) \cdot b/(b, c), (b, c) \cdot c/(b, c)] = (b, c)[b/(b, c), c/(b, c)]$. But $(b, c)[b/(b, c), c/(b, c)] = [b, c]$, so that $[b, c]/(b, c) = [b/(b, c), c/(b, c)]$; and so it remains for us to prove that $[b/(b, c), c/(b, c)] = b/(b, c) \cdot c/(b, c)$. I claim that the first stage of our two-stage process is complete, because $(b/(b, c), c/(b, c)) = 1$. It suffices for us to prove that the LCM of two relatively prime positive integers is equal to the product of those same two relatively prime positive integers. If r and s denote two relatively prime positive integers, we will be finished with this proof once we establish that $[r, s] = rs$. Now $s|[r, s] = tr$ for some positive integer t and $(s, r) = 1$, so $s|t$. So, if we write $t = sT$, then $[r, s] = rsT$. In order for $[r, s]$ to satisfy condition (b) of its definition on page 11, however, $T = 1$ is forced; and so (finally) $[r, s] = rs$ is established and, with it, our theorem.

Remark: Once we have the ideas of prime and of the so-called *canonical decomposition* (some prefer the term *unique factorization*) of positive integers, there will be other, easier ways of establishing results like the one we just struggled through. This fact can be regarded as substantial motivation for Chapter 2.

PROBLEMS

Solutions (or hints at solutions) to the problems marked with an asterisk (*) are given on page 121.

***1.** Find the least common multiple of b and c for the following pairs of positive integers:
(a) $b = 288$, $c = 51$.
(b) $b = 1{,}687$, $c = 482$.
***(c)** $b = 3{,}587$, $c = 1{,}819$.
(d) $b = 144$, $c = 89$.
[HINT (strong!): Use the information you got by doing problem 2 on page 5, and then apply Theorem 1.2.2 (page 11).]

***2.** In each of the following, decide whether there exist integers (not necessarily positive!) x, y, and z for which the equation holds. If you decide that the given equation is solvable in integers, then find one (or more?) solution(s). If you decide that the given equation has no integer solutions, then give a reason for your decision.
***(a)** $2x - 6y = 17$.
***(b)** $2x - 5y = 17$.
(c) $2x + 3y + 4z = -1$.

3. The GCD idea can easily be generalized to triples[1]—rather than pairs—of positive integers. If b, c, and d are positive integers, we define $G = (b, c, d) =$ the greatest common divisor of b, c, and d to be the positive integer satisfying: (i) $G|b$, $G|c$, and $G|d$; and (ii) if e is any integer for which $e|b$, $e|c$, and $e|d$, then $e|G$.
   (a) Find (2, 3, 4).
   (b) Is it correct to claim that $(b, c, d) = 1$ always forces $(b, c) = (b, d) = (c, d) = 1$? Prove that your answer is correct.
   (c) Is it correct to claim that, if the three positive integers b, c, and d are **relatively prime in pairs**—i.e., if $(b, c) = (b, d) = (c, d) = 1$—then $(b, c, d) = 1$ is forced? Prove that your answer is correct.

*4. If b, c, and d are positive integers, we define $L =$ the least common multiple of b, c, and d to be that positive integer, denoted $[b, c, d]$, satisfying: (i) $b|L$, $c|L$, and $d|L$; and (ii) if e is any integer for which $b|e$, $c|e$, and $d|e$, then $L|e$.
   (a) Find [2, 3, 4].
   *(b) Is $(b, c, d)[b, c, d] = bcd$ true for all positive integers b, c, and d? Prove that your answer is correct.

*5. Do there exist integers x and y satisfying the two conditions $x + y = 60$ and $(x, y) = 8$? Prove that your answer is correct.

*6. Do there exist integers x and y satisfying the two conditions $x + y = 60$ and $(x, y) = 12$? Prove that your answer is correct.

*7. Prove that there are infinitely many pairs of integers x and y satisfying the two conditions $x + y = 91$ and $(x, y) = 13$.

*8. Assume that G is a positive integer and that n is an integer. Prove that there exist integers x and y satisfying the two conditions $x + y = n$ and $(x, y) = G$ if and only if $G|n$. How many such solutions are there in case $G|n$? Proof?

*9. Assume that G and L are positive integers. Prove that there exist integers x and y satisfying the two conditions $(x, y) = G$ and $[x, y] = L$ if and only if $G|L$.

*10. Assume that G is a positive integer and that n is an integer. Prove that there exist integers x and y satisfying the two conditions $(x, y) = G$ and $xy = n$ if and only if $G^2|n$.

11. Prove the existence and uniqueness of the LCM, $[b, c]$, of any pair b and c of positive integers.

12. Prove that r_k satisfies condition (b) of the definition of GCD (see page 8).

*13. Establish properties 2 through 5 of the GCD, as listed on page 10.

[1]Or, more generally, n-tuples for any positive integer n.

14. Establish properties 1 and 2 of the LCM, as listed on page 11.

*15. Define $G = G(b, c)$ to be the GCD of the two integers $1 + b + b^2$ and $1 + b^c$, where b and c are arbitrary nonnegative integers. Find a formula for $G(b, c)$. [HINT (strong!): Think about geometric progressions and integral linear combinations.]

16.(a) Is it correct to claim that $(F_k, F_{k+1}) = 1$, regardless of the choice of the nonnegative integer k? Give a reason for your answer. [HINT: Refer back to page 6, problem 9, for the definition of the Fibonacci sequence if necessary.]

*(b) Is it correct to claim that $(F_m, F_n) = 1$, regardless of the choice of the ordered pair (m, n) of nonnegative integers m and n? Prove that your answer is correct.

*(c)†Obtain a formula for (F_m, F_n), valid regardless of the choice of the ordered pair (m, n) of nonnegative integers m and n. Prove that the formula you have obtained is correct.

17. Assume that $b > 1$ is an integer. Prove that $(b^m - 1, b^n - 1) = b^{(m,n)} - 1$, regardless of the choice of the ordered pair (m, n) of positive integers m and n. [HINT: Compare/contrast page 7, problem 13.]

18. (a) Prove that $G = (b, c)$ is, in fact, the greatest among all the common divisors of b and c in the sense that $d \leqslant G$ regardless of the choice of the common divisor d (of b and c).

(b) Evaluate (288, 51) and (144, 89) in the spirit of problem 18(a); i.e., for instance, list all divisors of 288, list all divisors of 51, list all common divisors of 288 and 51, and finally identify the greatest among all such common divisors.

19. (a) Is it correct to claim that, if $b|cd$, then necessarily $b|c$ or $b|d$? Prove that your answer is correct.

(b) Is there some distinguished subset of the set of positive integers—call it P—such that, if p is any element of P and $p|cd$, then necessarily $p|c$ or $p|d$? If your answer is yes, try to describe the set P as accurately as possible.

NOTES TO CHAPTER 1

1. The Euclidean Algorithm, in spite of its antiquity, is a pretty efficient way of determining the greatest common divisor of a pair of integers. For more precise information, see problem 12 on page 35 of LeVeque's *Fundamentals of Number Theory*. It is also perhaps interesting to know that The Euclidean Algorithm process is never less efficient than when trying to evaluate the GCD of two consecutive Fibonacci numbers.

2. Problem 8 on page 6 can be reinterpreted to say that the number e satisfies no equation of the form $f(X) = b_1X^1 + b_0 = 0$, regardless of the choice of the integers b_0 and b_1. It has been proved that e satisfies no polynomial equation with integral coefficients, regardless of the degree of the polynomial—i.e., the number e is **transcendental**—but the proof is difficult.
3. Birkhoff and MacLane's *A Survey of Modern Algebra* (revised edition), particularly on page 352, indicates that the GCD and LCM of two positive integers are their greatest lower bound and least upper bound under the divisibility relation—**dual** notions in an appropriate lattice.
4. Problem 2(a), (b), and (c) (page 12) are examples of **linear Diophantine equations:** equations in which each variable appears only to the first power (the linearity) and in which we are interested only in solutions that are integers.

2 Divisibility with Primes

INTRODUCTION

The most important thing in this chapter is the idea of what it takes for a positive integer to be a **prime.** We shall prove the **Fundamental Theorem of Arithmetic,** which says (roughly speaking) that every positive integer greater than 1 can be written as a product of primes and, further, that there is essentially only one way to write a given positive integer greater than 1 as a product of primes. This theorem provides us with a powerful and natural tool for solving problems, proving additional theorems, etc. In particular, we can learn a lot more about divisibility, greatest common divisors, and least common multiples by looking at these concepts from the point of view of the Fundamental Theorem of Arithmetic.

Thus, Chapter 2 should serve as a substantial review of Chapter 1, thereby leading to a deeper understanding of that material. I think that it is extremely important to give some examples of contexts in which there is no Fundamental Theorem; these examples will suggest the sorts of difficulties that can crop up when you don't have unique factorization and thus will make us grateful for the fact that we *do* have unique factorization in the context of the positive integers. In Section 2.3 a technique is introduced that enables us to focus our attention on an integer or rational number, one prime at a time. This technique allows us to view the concept of divisibility in a novel and, ultimately, fruitful way.

2.1 Primes and the Fundamental Theorem of Arithmetic

Definition 2.1.1: A positive integer $p > 1$ is said to be a **prime** if the only positive integral divisors of p are 1 and p.

Since every positive integer n has the property that $1|n$ and $n|n$, we see that being a prime amounts to having the absolute minimum number of positive integral divisors that the law allows. The integers 1 and n are sometimes referred to as the trivial positive integral divisors of the positive integer n. A positive integer $n > 1$ such as 6 fails to be prime—or is **composite**—by virtue of having at least one nontrivial divisor. In this case, 2 and 3 both serve as nontrivial divisors of 6.

The only stepping stone we need in order to be able to prove the central result of this chapter—the **Fundamental Theorem of Arithmetic**—is the following property.

***Property* (p):** If p is a prime such that $p|bc$, where b and c are integers, then either $p|b$ or $p|c$.[1]

Proof: We make our first application of the extremely useful (as we are already beginning to see!) Theorem 1.2.1 on page 10. We are given that our prime p divides the product of the integers b and c, and we need to show that p divides at least one of the pair b and c. If $p|b$, then we are done; so let's instead assume without of loss of generality that $p \nmid b$. Here is where the assumption that p is prime is brought to bear. Consider (p, b): (p, b) is a positive integral divisor of the prime p and therefore must be either 1 or p. Now $(p, b) \neq p$ because $p = (p, b)$ implies that $p|b$, which contradicts our without-loss-of-generality assumption that $p \nmid b$. So $(p, b) = 1$. But now we have $p|bc$ and $(p, b) = 1$, and so we can conclude via Theorem 1.2.1 that $p|c$. Thus in every case the prime p divides either b or c, and so the proof is complete.

Remarks:

1. $p|bc$ with p prime certainly doesn't enable one to conclude that both $p|b$ and $p|c$ hold, although this may be the case.

[1]Look back at page 14, problem 19(b), where the set P is the set of primes.

2. More significantly, by far, the property of primes just established *characterizes* primes among the positive integers; i.e., if n is composite, then n doesn't satisfy the above property. To illustrate, $4|bc$ doesn't force $4|b$ or $4|c$, as can be seen by choosing $b = 6$ and $c = 10$ or, in full generality, $b = 2r$ and $c = 2s$ for any choice of odd integers r and s. So Property (p) captures the very essence of "primeness" and could be used as a definition of prime. We shall encounter several other ways in which primes can be characterized later in the text.

We are now ready to state and prove the central result of this chapter, Theorem 2.1.1.

Theorem 2.1.1 – The Fundamental Theorem of Arithmetic:[1] Every integer $n > 1$ can be written **uniquely** in the form

$$n = p_1^{e_1} \cdot p_2^{e_2} \cdot \ . \ . \ . \ \cdot p_k^{e_k} \tag{1}$$

where $p_1 < p_2 < \ . \ . \ . \ < p_k$; each p_i, $1 \leqslant i \leqslant k$, is a prime; and each e_i, $1 \leqslant i \leqslant k$, is a positive integer.

Proof: We start by showing that every such n has at least one representation of the type described—this constitutes the existence part of the theorem—and then we prove the uniqueness of the representation. If n happens to be prime, say $n = p$, then $n = p = p_1^{e_1} \cdot p_2^{e_2} \cdot \ . \ . \ . \ \cdot p_k^{e_k}$ for $p_1 = p$, $e_1 = 1$, and $k = 1$, so that n has at least one representation of the type described. If n is composite, then (by the very definition of *composite*) n has a nontrivial divisor, say n_1, so that $1 < n_1 < n$ and yet $n_1|n$. Thus there exists a uniquely determined integer, say n_2, such that $n = n_1n_2$; and notice that $1 < n_2 < n$. If both n_1 and n_2 are primes, we have a representation of n of the desired type. Otherwise, we iterate the process on one or both of n_1 and n_2. For instance, suppose n_1 is prime but n_2 is composite. Then there is nothing more to be done with n_1, but $n_2 = n_3n_4$ for suitable integers n_3 and n_4 satisfying $1 < n_3 \leqslant n_4 < n_2$, so that now $n = n_1n_3n_4$. Since the size of the factor is strictly decreasing at each stage of the process, the process necessarily terminates after a finite number of steps (why?), at which time we have a representation of the desired type for n.

So much for the argument establishing the existence of at least one representation of the requisite type for each integer $n > 1$. Now for the uniqueness part of the theorem. We make a proof by contradiction, as follows. Suppose that there is an integer $n > 1$ for which there exist two distinct representations of the type described in the statement of the theorem. Then we would have $n = p_1^{e_1} \cdot p_2^{e_2} \cdot$

[1] Also referred to as the Unique Factorization Theorem or the Canonical Decomposition Theorem.

$\cdots p_k^{e_k} = q_1^{f_1} \cdot q_2^{f_2} \cdot \ldots \cdot q_r^{f_r}$. It might be the case that some primes are to be found in both these representations of n. If so, divide each side of the equation by all such common primes, thereby getting an equation of the following type: $P_1^{E_1} \cdot \ldots \cdot P_g^{E_g} = Q_1^{F_1} \cdot \ldots \cdot Q_s^{F_s}$, where each prime P_i is one of the original set $\{p_1, \ldots, p_k\}$ and where likewise each prime Q_j comes from the set $\{q_1, \ldots, q_r\}$, but now $P_i \neq Q_j$ regardless of the choice of the subscripts i and j. However, $P_1 | P_1^{E_1} | P_1^{E_1} \cdot \ldots \cdot P_g^{E_g} = Q_1^{F_1} \cdot \ldots \cdot Q_s^{F_s}$, in conjunction with Property (p), tells us that $P_1 | Q_j$ for some subscript j and hence $P_1 = Q_j$ for that subscript j, because P_1 and Q_j are both primes. However, this is now a contradiction because $P_i \neq Q_j$ regardless of the choice of the subscripts i and j, and so the proof is complete.

Example 2.1.1: $60 = 2^2 3^1 5^1$, $17 = 17^1$, $210 = 2^1 3^1 5^1 7^1$, and $1{,}093 = 1{,}093^1$ because 1,093 is a prime (how do you know?).

As a first illustration of the power and "naturalness" of the Fundamental Theorem of Arithmetic, we shall give a second proof of the result (see Theorem 1.2.2 on page 11) that the product of the GCD and the LCM of two positive integers is equal to the product of the two positive integers themselves. In order to streamline the argument, we revamp the canonical decomposition notation slightly, as follows. Let $\{p_1, p_2, \ldots, p_k\}$ be the set-theoretic union of the set of distinct primes dividing b and the set of distinct primes dividing c. Then we may write the canonical decomposition of b in the form $b = p_1^{e_1} \cdot p_2^{e_2} \cdot \ldots \cdot p_k^{e_k}$—with the understanding that, if the prime p_i doesn't divide b, then the exponent e_i is zero. Likewise, we write $c = p_1^{f_1} \cdot p_2^{f_2} \cdot \ldots \cdot p_k^{f_k}$.

Now consider the positive integer $G = p_1^{\min\{e_1,f_1\}} \cdot p_2^{\min\{e_2,f_2\}} \cdot \ldots \cdot p_k^{\min\{e_k,f_k\}}$, where $\min\{e_i, f_i\}$ denotes the smaller of the two exponents e_i, f_i. I claim that the positive integer G just described is (b, c), the GCD of b and c. To see, for instance, that $G|b$, write $b = G \cdot p_1^{e_1 - \min\{e_1,f_1\}} \cdot (p_2^{e_2 - \min\{e_2,f_2\}} \cdot \ldots \cdot p_k^{e_k - \min\{e_k,f_k\}})$; notice that this second factor is an integer, because each exponent is of the form $e_i - \min\{e_i, f_i\}$, and hence is necessarily a nonnegative integer. Likewise, $G|c$, and hence G is (at least) a common divisor of b and c. On the other hand, any positive integer capable of serving as a common divisor of b and c must be of the form $p_1^{g_1} \cdot p_2^{g_2} \cdot \ldots \cdot p_k^{g_k}$, where $g_i \leq \min\{e_i, f_i\}$ for every value of i. Suppose, for instance, that $g_1 > \min\{e_1, f_1\}$. Then either $g_1 > e_1$ or $g_1 > f_1$. If $g_1 > e_1$, then $p_1^{g_1} \cdot p_2^{g_2} \cdot \ldots \cdot p_k^{g_k} \nmid b$ (why?). If $g_1 > f_1$, then $p_1^{f_1} \cdot p_2^{f_2} \cdot \ldots \cdot p_k^{f_k} \nmid c$. It thus follows that G is the greatest among the positive integral divisors of b and c. Likewise, $p_1^{\max\{e_1,f_1\}} \cdot p_2^{\max\{e_2,f_2\}} \cdot \ldots \cdot p_k^{\max\{e_k,f_k\}}$, ($\max\{e_i, f_i\}$ denoting the larger of the two exponents e_i and f_i) represents the least common multiple of b and c in the form used in canonical decomposition. Finally,

$$\begin{aligned}(b, c)[b, c] &= (p_1^{\min\{e_1,f_1\}} \cdot \ldots \cdot p_k^{\min\{e_k,f_k\}}) \cdot (p_1^{\max\{e_1,f_1\}} \cdot \ldots \cdot p_k^{\max\{e_k,f_k\}}) \\ &= p_1^{\min\{e_1,f_1\} + \max\{e_1,f_1\}} \cdot \ldots \cdot p_k^{\min\{e_k,f_k\} + \max\{e_k,f_k\}} \\ &= p_1^{e_1 + f_1} \cdot \ldots \cdot p_k^{e_k + f_k} = (p_1^{e_1} \cdot \ldots \cdot p_k^{e_k}) \cdot \ldots \cdot (p_1^{f_1} \cdot \ldots \cdot p_k^{f_k}) = bc,\end{aligned}$$

as desired, where we have used the laws of exponents and the fact that min $\{r, s\}$ + max $\{r, s\} = r + s$, regardless of the choice of the real numbers r and s.

The argument above is pretty long ("unduly protracted" might be a more accurate assessment) and so may not have convinced you of "the power and 'naturalness' of the Fundamental Theorem of Arithmetic." But the argument is long only because I put in every last detail. Analyzing the nature of the argument—looking beyond the formal symbolism used—is extremely important. Everything hinges on the fact that, if the canonical decompositions of m and n are $m = p_1^{m_1} \cdot p_2^{m_2} \cdot \ldots \cdot p_k^{m_k}$ and $n = p_1^{n_1} \cdot p_2^{n_2} \cdot \ldots \cdot p_k^{n_k}$, then $m|n$ if and only if $m_i \leqslant n_i$ for all i such that $1 \leqslant i \leqslant k$. The technique just illustrated makes it fairly straightforward to establish results like $(b, c, d) = ((b, c), d)$ (see problem 10(a) on page 23), as well as the somewhat more exotic results $|bcd| = [b, c, d](bc, cd, db)$ and, dually, $(b, c, d)[bc, cd, db] = |bcd|$ (see problem 10(b) and (c) on page 23).

We finish the body of this section by stating and proving the famous theorem of Euclid on the infinitude of primes.

Theorem 2.1.2 (Euclid):[1] There are infinitely many primes.

Proof: We make the proof by contradiction; i.e., we start by assuming that there are only finitely many primes. In that case we may claim, notationally, that the finite set $\{p_1, p_2, \ldots, p_k\}$ is a listing of the entire set of primes. To get our desired contradiction, it is enough to argue that at least one prime p isn't in this list. Euclid's argument goes as follows: consider the integer $n > 1$ defined by $n = (p_1 p_2 \cdot \ldots \cdot p_k) + 1$. By the Fundamental Theorem of Arithmetic, n has its canonical decomposition; therefore, in particular, there must exist at least one prime p appearing in n's canonical decomposition and of course $p|n$. We will be done once we prove that $p \neq p_i$, regardless of the choice of i with $1 \leqslant i \leqslant k$. What would happen if $p = p_i$ for some i satisfying $1 \leqslant i \leqslant k$?[2] We would have $p = p_i | p_1 \cdot p_2 \cdot \ldots \cdot p_{i-1} \cdot p_i \cdot p_{i+1} \cdot \ldots \cdot p_k = n - 1$ and also $p|n$, and hence (why?) $p|1$, which is our contradiction (again, why?).

PROBLEMS

Solutions (or hints at solutions) to the problems marked with an asterisk (*) are given on page 122.

*1. (a) If n is composite (by virtue of having at least one nontrivial positive

[1]Proposition 20, Book IX.

[2]Here we have a proof by contradiction nested inside a proof by contradiction.

integral divisor), does it follow that n must have at least two nontrivial positive integral divisors? Prove that your answer is correct.

*(b) Characterize the positive integers n that have precisely one nontrivial positive integral divisor; i.e., the positive integer n has exactly one nontrivial positive integral divisor if and only if—what?

*2. (a) Make a list of the primes less than 100.

*(b) Prove that, if the positive integer n has no prime divisor p satisfying the inequality $p \leqslant \sqrt{n}$, then n is prime.

*3. On page 20 we found formulae for $G = (b, c)$ and $L = [b, c]$ in terms of the **canonical decompositions** of the individual positive integers b and c. Use these formulae to recompute (see page 5, problem 2, and page 12, problem 1) the greatest common divisor and least common multiple of the following pairs of positive integers:

*(a) $b = 288$, $c = 51$.
(b) $b = 1{,}687$, $c = 482$.
(c) $b = 3{,}587$, $c = 1{,}819$.
(d) $b = 144$, $c = 89$.

4. Let's define, for each positive integer n, the positive integer G_n via the formula $G_n = 2^{2^n} - 1$.
(a) Evaluate G_1, G_2, and G_3.
(b) Prove that $G_n | G_{n+1}$ for every positive integer n.
(c) Prove that G_n has at least n distinct prime factors for every positive integer n.
(d) Give another proof of the fact that there are infinitely many primes.

5. Let's define, for each nonnegative integer n, the positive integer f_n via the formula $f_n = 2^{2^n} + 1$.
(a) Evaluate f_0, f_1, f_2, and f_3.
(b) Prove that $(f_n, f_{n+1}) = 1$ for every positive integer n.
(c) Give yet another proof of the fact that there are infinitely many primes.

*6. Assume that b, m, and n are positive integers.

*(a) Prove that, if $n < m$, then $(b^{2^n} + 1) | (b^{2^m} - 1)$.

*(b) Prove that, if $n \neq m$, then $(b^{2^n} + 1, b^{2^m} + 1) = 1$ if b is even and $(b^{2^n} + 1, b^{2^m} + 1) = 2$ if b is odd.

7. The number 1 was not allowed to be called a prime because $p > 1$ was legislated in Definition 2.1.1. On the other hand, the number 1 satisfies the primeness requirement of having only trivial positive integral divisors. So why isn't it a good idea to let 1 be a prime? [HINT: Think about the Fundamental Theorem of Arithmetic.]

*8. We define the **number of divisors function** τ as follows: for every positive integer n, $\tau(n) =$ the total number of positive integral divisors of n. So $\tau(1) = 1$, $\tau(2) = 2$, $\tau(3) = 2$, $\tau(4) = 3$, $\tau(16) = 5$, $\tau(24) = 8$, etc.

***(a)** Characterize primes via the τ function; i.e., p is a prime if and only if $\tau(p) =$ what?

***(b)** Characterize the positive integers n for which $\tau(n)$ is odd; i.e., $\tau(n)$ is odd if and only if what? [HINT: Refer to problem 4 on page 5.]

***(c)** If $n = p^e$, where p is any prime and e is any positive integer, find a formula for $\tau(n)$.

(d) Assuming that $(n, m) = 1$ implies $\tau(nm) = \tau(n)\tau(m)$, obtain a formula for $\tau(n)$ for every positive integer n. [HINT: Use part (c) of this problem, in conjunction with the Fundamental Theorem of Arithmetic.]

†(e) Is the assumption made in part (d) of this problem correct? Prove that your answer is correct.

9. **The product notation, the *p*-components of a positive integer *n*, and divisibility.** We have been writing the canonical decomposition of n in the notation $n = p_1^{e_1} \cdot p_2^{e_2} \cdot \ldots \cdot p_k^{e_k}$. The product notation way of expressing this equation is

$$n = \prod_{i=1}^{k} p_i^{e_i}$$

or, alternatively,

$$n = \prod_{\substack{p \text{ prime} \\ p|n}} p^e .$$

Here Π is the capital Greek letter pi, the multiplicative analogue of $\sum$ (capital sigma), the more familiar summation notation. We refer to p^e as the ***p*-component** of n, following LeVeque in *Fundamentals of Number Theory*. Prove that $b|c$ if and only if every p-component of b is less than or equal to the corresponding p-component of c.

10. (a) Prove that $(b, c, d) = ((b, c), d) = ((c, d), b) = ((b, d), c)$.
(b) Prove that $[b, c, d](bc, cd, db) = |bcd|$.
(c) Prove that $(b, c, d)[bc, cd, db] = |bcd|$. [HINT (strong!): Use the Fundamental Theorem of Arithmetic.]
†(d) The result in part (c) of this problem could be obtained formally from the result in part (b), by systematically interchanging GCD brackets and LCM brackets. Can this formal manipulation be justified mathematically?

***11.** In each case decide whether the statement is true or false, and give a proof or counterexample.
***(a)** If p is a prime and $p|(b^2 + c^2)$ and $p|(c^2 + d^2)$, then $p|(b^2 + d^2)$.
***(b)** If $M_p = 2^p - 1$ is a prime, then p must be a prime.
***(c)** If p is a prime and $p|(b^2 - c^2)$ and $p|(c^2 - d^2)$, then $p|(b^2 - d^2)$.
***(d)** If p is a prime, then $M_p = 2^p - 1$ is a prime.

***(e)** If p is a prime and $p|(b^2 + c^2)$ and $p|(c^2 + d^2)$, then $p|(b^2 - d^2)$.
***(f)** If p is a prime and $p|b$ and $p|(b^2 + c^2)$, then $p|c$.

12. (a) Prove that there are infinitely many primes p of the form $p = 4n + 3$, where n is a positive integer.
(b) Prove that there are infinitely many primes in the arithmetic progression $\{5, 11, 17, 23, 29, 35, \ldots\}$.
[HINT (strong!): For part (a), dream up an n that is like the $n = p_1 \cdot p_2 \cdot \ldots \cdot p_k + 1$ of Euclid's proof, but force your n to be of the form $n = 4m + 3$ for some suitable positive integer m.]

†13. Assume that b and c are positive integers and that $p^e \| (b^c - 1)$, where p is an odd prime and $e > 0$. Prove that $p^e \| c(b^{(c,p-1)} - 1)$. [Here the notation $p^e \| (b^c - 1)$ means that $p^e | (b^c - 1)$ but that $p^{e+1} \nmid (b^c - 1)$.]

***14.** Prove that the product of any k consecutive integers is divisible by $k!$

***15.** Determine all those positive integers b and c with $b < c$ for which $b^c = c^b$.

***16.** **Arithmetic, geometric, and harmonic numbers.** We first define the **sum of divisors function** σ as follows: for every positive integer n, $\sigma(n) =$ the sum of the positive integral divisors of n. So $\sigma(1) = 1$, $\sigma(2) = 3$, $\sigma(4) = 7$, $\sigma(25) = 31$, etc.

***(a)** If $n = p^e$, where p is any prime and e is any positive integer, find a formula for $\sigma(n)$.

Let us denote by $S(n)$ the set of positive integral divisors of n. We now define $A(n)$ to be the **arithmetic mean** of the elements of the set $S(n)$—i.e., the sum of the elements of the set $S(n)$ divided by the number of elements of the set $S(n)$. For instance, $A(1) = 1/1 = 1$, $A(2) = (1 + 2)/2 = 3/2$, $A(4) = (1 + 2 + 4)/3 = 7/3$, $A(5^3) = (1 + 5 + 5^2 + 5^3)/4 = 39$, $A(5^5) = (1 + 5 + 5^2 + 5^3 + 5^4 + 5^5)/6 = 651$, etc. A positive integer n is said to be an **arithmetic number** if the (at worst, rational) number $A(n)$ is an integer.

(b) The *arithmetic mean* of two numbers b and c is defined to be $(b + c)/2$. Prove that $A(n)$ specializes to the arithmetic mean if and only if n is a prime.
***(c)** Are there infinitely many arithmetic numbers? Proof?
(d) Obtain a formula for $A(n)$ in terms of $\tau(n)$ and $\sigma(n)$.
(e) Obtain a formula for $A(p^e)$, where p is any prime and e is any positive integer.
(f) Is it correct to claim that $A(p^e)$ is integral infinitely often if we require $e > 1$? Proof?

Just as $A(n)$ is based on the arithmetic mean, $G(n)$ and $H(n)$ are based, respectively, on the **geometric mean** and the **harmonic mean.** The *geometric mean* of two nonnegative real numbers b and c is defined to be $\sqrt{bc}$, while the *harmonic mean* of the two positive real numbers b and c is defined to be $2/(1/b + 1/c)$. We define $G(n)$ to be the geometric mean of the elements of the set $S(n)$, introduced previously in this problem. $G(4) = (1 \cdot 2 \cdot 4)^{1/3}$, for instance, since $S(4) = \{1, 2, 4\}$ and $\#(S(n)) = 3$. $H(4)$ would be $3/(1/1 + 1/2 + 1/4)$, the harmonic mean of the elements of the set $S(4)$. In general, $H(n)$ is defined to be the harmonic mean of the elements of the set $S(n)$.

(g) Get a formula for $G(n)$ in terms of $\tau(n)$ and $\sigma(n)$.

(h) Get a formula for $G(n)$ that doesn't involve either of the functions τ and σ.

***(i)** Characterize the **geometric numbers**—i.e., the positive integers n for which $G(n)$ is an integer.

(j) Get a formula for $H(n)$ in terms of n, $\tau(n)$, and $\sigma(n)$.

(k) Prove that no pure prime power can be a **harmonic number;** i.e., $H(p^e)$ isn't an integer, regardless of the choice of the prime p and the positive integer e.

(l) A positive integer n is said to be a **perfect number** if $\sigma(n) = 2n$. Prove that every even perfect number is a harmonic number.

†(m) If n is a harmonic number, must it be even?

17. Find all positive integral solutions (if any) of the equation $(2^x - 1)(2^y - 1) = 2^{2^z} - 1$.

***18.** Assume that b and c are integers such that $b + c \neq 0$ and assume that p is an odd prime. Prove that $(b + c, (b^p + c^p)/(b + c))$ is either 1 or p.

2.2 Examples of Contexts *Not* Possessing Unique Factorization

The object of this section is to provide some examples of contexts in which there is no "Fundamental Theorem." It is instructive to compare and contrast these "pathological" contexts with our usual context of the (positive) integers. We give two examples of such "pathological" contexts: the first example is simpler and easier to describe (which is why it is the first example!) than the second, but the second example is much more important for advanced work.

Example 2.2.1: Our viewpoint here is that our entire universe is composed of the set of even positive integers $E = \{2, 4, 6, 8, \ldots\}$. Let us say that an element e of the set E is **indecomposable**[1] (or E-indecomposable, if we want to be explicit

[1] I prefer to use the term *indecomposable* rather than *prime* because the set of numbers we are working with here lacks unique factorization.

about our context) if and only if e cannot be written as a product of two elements of E. So 6, for instance, qualifies as an E-indecomposable element because—although $6 = 2 \cdot 3$ is composite in our standard context—3 no longer exists, and hence 6 doesn't factor in E. It is easy to check that the first few E-indecomposables are 2, 6, 10, 14, 18, and 22. This pattern makes it extremely tempting to conjecture[1] that every second element of E is indecomposable, and in this case it is easy to verify this conjecture (see problem 2 on page 27). Assuming that this verification has been carried out, we would know, for instance, that 30 is an indecomposable element of E. Of course we could verify directly that 30 is E-indecomposable by noting that the only factorizations of 30 in the positive integers are $30 = 1 \cdot 30 = 2 \cdot 15 = 3 \cdot 10 = 5 \cdot 6$, none of which is a factorization within E. Now for the punch line: $60 = 2 \cdot 30 = 6 \cdot 10$ shows that the element 60 (of E) has two different "canonical decompositions" in E: we have already seen that each of 2, 6, 10, and 30 is an indecomposable element of E, so this equation shows that the number 60 can be written as a product of indecomposable elements of E in two different ways. Hence there can be no Fundamental Theorem of Arithmetic for the context E!

Example 2.2.2: Consider the set S of all complex numbers of the form $b + c\sqrt{-10}$, where b and c are arbitrary integers. Our aim is to show that S has no Fundamental Theorem of Arithmetic. Mimicking the definition of E-*indecomposable element* in Example 2.2.1, we say that an element of S is **indecomposable** (or, again, **S-indecomposable,** if we want to make our context explicit) if and only if $b + c\sqrt{-10}$ can't be written as a product of two elements of S, except for the two trivial factorizations $b + c\sqrt{-10} = 1 \cdot (b + c\sqrt{-10})$ or $b + c\sqrt{-10} = (-1) \cdot (-b - c\sqrt{-10})$. The equation $14 = 2 \cdot 7 = (2 + \sqrt{-10}) \cdot (2 - \sqrt{-10})$ will allow us to conclude that our context S doesn't possess the unique factorization property, provided we can show that each of 2, 7, $2 + \sqrt{-10}$, and $2 - \sqrt{-10}$ is S-indecomposable. (Please notice the parallelism between the argument here and that of Example 2.2.1.) The numbers 2 and 7 are indecomposable elements of S (proofs?). I will prove that $2 + \sqrt{-10}$ is S-indecomposable, and leave to the reader the job of proving that $2 - \sqrt{-10}$ is S-indecomposable. Suppose that $2 + \sqrt{-10} = (u + v\sqrt{-10}) \cdot (r + s\sqrt{-10})$ with u, v, r, and s integers. Then $2 - \sqrt{-10} = (u - v\sqrt{-10}) \cdot (r - s\sqrt{-10})$ (why?); and so, multiplying these two equations together, we have $14 = (u^2 + 10v^2) \cdot (r^2 + 10s^2)$. Now $u^2 + 10v^2$ and $r^2 + 10s^2$ are nonnegative integers, and there is no loss of generality in assuming that $u^2 + 10v^2 \leq r^2 + 10s^2$, so there are only two logical possibilities: $u^2 + 10v^2 = 1$, in which case $r^2 + 10s^2 = 14$ is forced; or $u^2 + 10v^2 = 2$, in which case $r^2 + 10s^2 = 7$ is forced. The equation $u^2 + 10v^2 = 2$ has no integral solutions,

[1]Rough definition: a mathematician's educated guess.

so our original assumption that $2 + \sqrt{-10} = (u + v\sqrt{-10}) \cdot (r + s\sqrt{-10})$ has forced the condition $u^2 + 10v^2 = 1$, which in turn forces $v = 0$, $u = \pm 1$, and thus finally $r + s\sqrt{-10} = \pm(2 + \sqrt{-10})$. The conclusion is that the only factorizations of $2 + \sqrt{-10}$ as a product of two elements of S are the two trivial S-factorizations $2 + \sqrt{-10} = 1 \cdot (2 + \sqrt{-10})$ and $2 + \sqrt{-10} = (-1) \cdot (-2 - \sqrt{-10})$, and so we have proved that $2 + \sqrt{-10}$ is S-indecomposable. The element $2 - \sqrt{-10}$ of S can be shown to be S-indecomposable element (see problem 4 on page 27), so now we have a counterexample to the claim that S has a Fundamental Theorem of Arithmetic: the equation $14 = 2 \cdot 7 = (2 + \sqrt{-10}) \cdot (2 - \sqrt{-10})$ tells us that the element 14 of the context S has two distinct factorizations into indecomposable elements of S.

PROBLEMS

Solutions (or hints at solutions) to the problems marked with an asterisk (*) are given on page 123.

***1.*(a)** If p is an indecomposable element of E and if $p|bc$, where b and c are both elements of E, does it necessarily follow that $p|b$ or $p|c$? Prove that your answer is correct. (Here the notation $b|c$ is to be interpreted as divisibility with respect to the context E; i.e., $b|c$ if and only if there exists a unique element d of E such that $c = bd$.)

(b) If $b + c\sqrt{-10}$ is an indecomposable element of S and $b + c\sqrt{-10}|(d + e\sqrt{-10}) \cdot (f + g\sqrt{-10})$, where both $d + e\sqrt{-10}$ and $f + g\sqrt{-10}$ are elements of S, does it follow that $b + c\sqrt{-10}|(d + e\sqrt{-10})$ or $b + c\sqrt{-10}|(f + g\sqrt{-10})$? Prove that your answer is correct. (Here the notation | represents divisibility in S, or S-divisibility, which is defined in precisely the same way as were divisibility (Z-divisibility, if you like) and E-divisibility.)

2. Prove that the indecomposable elements of E are precisely the numbers of the form $2n$, as n runs through the set of all odd positive integers.

***3.*(a)** If p is a prime, must p be an indecomposable element of E? Prove that your answer is correct.

***(b)** If p is an indecomposable element of E, must p be a prime? Prove that your answer is correct.

***(c)** If $b + c\sqrt{-10}$ is an indecomposable element of S, must it be a prime? Prove that your answer is correct.

(d) If p is a prime, must p be an indecomposable element of S? Prove that your answer is correct.

4. Prove that $2 - \sqrt{-10}$ is an indecomposable element of S.

5. Prove that infinitely many elements of E are like 60 in that they can be written as a product of indecomposable elements of E in more than one way.

6. Give an alternate proof of the fact that $2 + \sqrt{-10}$ is an indecomposable element of S, by equating real and imaginary parts in the equation $2 + \sqrt{-10} = (u + v\sqrt{-10}) \cdot (r + s\sqrt{-10}) = (u + (v \cdot 10i)) \cdot (r + (s \cdot 10i))$.

*7. Give an argument showing that the equation $2 - \sqrt{-10} = (u - v\sqrt{-10})(r - s\sqrt{-10})$ follows from the equation $2 + \sqrt{-10} = (u + v\sqrt{-10})(r + s\sqrt{-10})$.

*8. Let's define the **norm** of a generic element of S. The *norm* of $b + c\sqrt{-10}$, written $N(b + c\sqrt{-10})$, is defined to be the product of $b + c\sqrt{-10}$ by its **complex conjugate,** $b - c\sqrt{-10}$. Thus, $N(b + c\sqrt{-10}) = (b + c\sqrt{-10})(b - c\sqrt{-10}) = b^2 + 10c^2$. One important property of the norm is that it is a function from S into the nonnegative integers. Prove the most important property of the norm—the fact that it is **multiplicative,** i.e., $N((b + c\sqrt{-10})(d + e\sqrt{-10})) = N(b + c\sqrt{-10}) \cdot N(d + e\sqrt{-10})$.

9. Prove for each of the contexts E and S that, while we have lost the uniqueness property, it is still true that every element of E (respectively, S) is a product of indecomposable elements of E (respectively, S). [HINT: Use the multiplicative property of the norm (see problem 8) in your proof for the context S.]

10. Assuming that the infinite process involved makes sense, evaluate

$$\alpha = 1 + \sqrt{1 + \sqrt{1 + \sqrt{1 + \cdots}}}$$

*11. **Infinite simple continued fractions.**[1]
   (a) Assuming that the infinite process involved makes sense, evaluate $\beta = 1 + 1/(1 + 1/(1 + 1/(1 + \cdots$
   *(b) Compare the values of α, from problem 10, and β, from problem 11(a).
   *(c)† Does the set $Z + Z[2\beta] = \{m + n(2\beta)$: m and n each taking on all integral values$\}$ possess a Fundamental Theorem of Arithmetic? Proof?
   (d) Show that the "partial sums" $1 = 1/1$, $1 + 1/1 = 2/1$, $1 + 1/(1 + 1) = 1 + 1/2 = 3/2, \ldots$ of β reproduce the Fibonacci sequence (see page 6, problem 9).
   (e) Show that each of the two **Pell equations,** $X^2 - 5V^2 = \pm 4$, has infinitely many solutions, by considering even and odd positive integral powers of $(1 + \sqrt{5})/2$.

[1] See pages 6–7, problem 12 for **finite simple continued fractions.**

(f) Draw the **Klein diagram** of the number $(1 + \sqrt{5})/2$, by (i) sketching the graph of the straight line $y = ((1 + \sqrt{5})/2 \cdot x$, and (ii) plotting the lattice points (1, 1), (1, 2), (2, 3), (3, 5), ...

2.3 p-adic Valuations: A New Type of Absolute Value

The object of this final section of Chapter 2 is to introduce yet another way for us to decide whether or not one integer divides another. We already have the definition of divisibility itself and the canonical decomposition approach. One convenient way to advertise the canonical decomposition approach is to say that $b|c$ if and only if every p-component of b is less than or equal to the corresponding p-component of c(see problem 9 on page 23). Here, the p-component of b is the part of the canonical decomposition of b that involves the prime p. For example, the 2-component of 360 is $8 = 2^3$, and the 3-component of 360 is $9 = 3^2$, because the canonical decomposition of 360 is $360 = 2^3 \cdot 3^2 \cdot 5^1$.

Before giving the basic definition of this section, we need a little preparation. Let q be any rational number (see problem 8 on page 6 for the definition of rational number). Let p be any arbitrary but fixed prime. Then q can be written uniquely in the form $q = (b/c)p^n$, where b, c, and n are integers; c is positive; $(b, c) = 1$; and $p \nmid bc$ (prove this!).

Remark: What we have done here is to write the rational number q as a product of two (rational) numbers: the first factor b/c does not involve the preassigned prime p at all (and constitutes the "p-free" part of q), while the second factor p^n is totally involved with the prime p (and constitutes the "p-full" part of q).

We are now ready to introduce the basic concept of this section.

Definition 2.3.1: Let p be any arbitrary but fixed prime. For each nonzero rational number q, **the p-adic absolute value of q,** denoted $|q|_p$, is defined to be $|q|_p = p^{-n}$, where it is to be understood that q has been written in the form $q = (b/c)p^n$ (as described just prior to this definition). We make the special definition $|0|_p = 0$.

Example 2.3.1: Since $360 = 2^3 \cdot 3^2 \cdot 5^1$, $|360|_2 = 2^{-3} = 1/8$ because $q = (b/c)p^n$ becomes $360 = (3^2 \cdot 5^1) \cdot 2^3 = 45 \cdot 2^3$ so that $q = 360$, $b = 45$, $c = 1$, $p = 2$, and $n = 3$. Likewise, $|360|_3 = 1/9$, $|360|_5 = 1/5$, and $|360|_p = 1$ if the prime $p \neq 2, 3$, or 5. The ordinary absolute value of 360 is, of course,

$|360| = 360$. Notice that $|360|_p \leqslant 1$ for every prime p, and that the product of all the p-adic absolute values is exactly the reciprocal of $|360|$. This product could be written as $\prod_p |360|_p$, where we have made use of the product notation (see problem 9 on page 23) and where it is to be understood that p runs through the set of all primes. So $\prod_p |360|_p$, at first blush, looks like an infinite product (because we know, thanks to Euclid, that there are infinitely many primes); but it reduces to the finite product $|360|_2 \cdot |360|_3 \cdot |360|_5 = (1/2^3)(1/3^2)(1/5^1) = |360|^{-1}$, since $|360|_p = 1$ for every prime p other than 2, 3, or 5 because 360 has only the primes 2, 3, and 5 in its canonical decomposition.

Remark: How big is 360? That sounds like a silly question but only because we are slaves to the usual measuring stick, namely $|\cdot|$. Now that we have infinitely many measuring sticks—namely $|\cdot|, |\cdot|_2, |\cdot|_3, \ldots |\cdot|_p, \ldots$—the question has more content. For instance, the 3-adic size of 360 is $3^{-2} = 1/9$; so from this point of view, 360 doesn't look as big as it once did!

Our next job is to relate this somewhat bizarre way of measuring size to the central concepts of divisibility and canonical decomposition. In order to do this we need to establish the following property of the function $|\cdot|_p$.

Equation (1): $|qr|_p = |q|_p \cdot |r|_p$, regardless of the choice of the rational numbers q and r, and regardless of the choice of the prime p.

Remark: We say that the function $|\cdot|_p$ is **multiplicative.**[1]

Proof of Equation (1): Let $q = (b/c)p^n$ and $r = (s/t)p^m$, where $b, c, s, t, n,$ and m are integers; $b > 0$; $c > 0$; $(b, c) = (s, t) = 1$; and $p \nmid bcst$. Then $|q|_p = p^{-n}$ and $|r|_p = p^{-m}$, so that $|q|_p \cdot |r|_p = p^{-n} \cdot p^{-m} = p^{-(n+m)}$, and we will be done as soon as we show that $|qr|_p = p^{-(n+m)}$. Now $|qr|_p = |((b/c)p^n) \cdot ((s/t)p^m)|_p = |(bs/ct) \cdot p^{n+m}|_p = p^{-(n+m)}$, and we are finished with this proof.

Now for our first theorem.

Theorem 2.3.1: $b|c$ if and only if $|c|_p \leqslant |b|_p$, for every prime p.

Proof: First, $b|c$ if and only if c/b is an integer. Second, c/b is an integer if and only if $|c/b|_p \leqslant 1$ for all primes p (why?). And finally, the requirement $|c/b|_p \leqslant 1$ for all primes p is equivalent to insisting that $|c|_p \leqslant |b|_p$ for all primes p because of the multiplicativity of the function $|\cdot|_p$ (see Equation (1)).

[1]See page 78, note 6.

Remark: This little result shows the intimate relationship between the two concepts of divisibility and p-adic valuations.

Example 2.3.2: (a) $24|120$ because $|24|_2 = 2^{-3} = 1/8 = |120|_2$, $|120|_3 = 3^{-1} = |24|_3$, $|120|_5 = 5^{-1} \leqslant |24|_5$, and $|120|_p = |24|_p$ for all primes p other than 2, 3, or 5. (b) $105 \nmid 1{,}001$ because $|1{,}001|_3 = 1 > 3^{-1} = |105|_3$.

Remark: The astute reader will notice (has already noticed?) that, if b really does divide c (as in the case of $24|120$ in Example 2.3.2(a)), then to verify this p-adically requires that we know the canonical decomposition of both b and c. So why bother with the p-adic approach at all? The answer is that, although the p-adic approach may be regarded as "equivalent" to canonical decomposition in some sense, it is often more convenient to work p-adically. More importantly, the p-adic approach can be developed a great deal further and ultimately becomes an extremely important way of attacking problems in mathematics generally—not just in number theory. Problems 5 and 6 at the end of this section (pages 32–33) are designed to give the reader a hint of how things p-adic can be related to standard concepts from calculus (sequences and series) and to geometry.

We finish our brief introduction to p-adic valuations of the field[1] of rational numbers by establishing an extremely important inequality that has a fancy-sounding name.

Theorem 2.3.2 – The Ultrametric Inequality: For any arbitrary but fixed prime p, we have the inequality $|q + r|_p \leqslant \max\{|q|_p, |r|_p\}$ for every choice of the rational numbers q and r.

Proof: We use the same notation that we used in proving that the function $|\cdot|_p$ is multiplicative: $q = (b/c)p^n$ and $r = (s/t)p^m$, with b, c, n, m, s, and t integers; $(b, c) = (s, t) = 1$; $c > 0$; $t > 0$; and $p \nmid bcst$. We may also assume, without loss of generality, that $n \geqslant m$; therefore, $|q|_p = p^{-n} \leqslant p^{-m} = |r|_p$, which means that we need to show that $|q + r|_p \leqslant |r|_p = p^{-m} = \max\{|q|_p, |r|_p\}$. The rest of the proof consists of standard algebraic manipulations: $q + r = (b/c)p^n + (s/t)p^m = (p^m/ct)(btp^{n-m} + sc)$, and so $|q + r|_p = |(p^m/ct)(btp^{n-m} + sc)|_p = |p^m/ct|_p \cdot |btp^{n-m} + sc|_p$ (using the multiplicative property of $|\cdot|_p$) $= p^{-m} \cdot |btp^{n-m} + sc|_p$, where we have used the assumption $p \nmid bcst$ to conclude that $p \nmid ct$. Since $n \geqslant m$, $n - m \geqslant 0$; thus $btp^{n-m} + sc$ is an integer, and hence $|btp^{n-m} + sc|_p \leqslant 1$, regardless of the prime p (refer to the end of the proof of Theorem 2.3.1 on page 30). Putting everything together, we have $|q + r|_p = p^{-m}|btp^{n-m} + sc|_p \leqslant p^{-m} = \max\{|q|_p, |r|_p\}$, and we are done.

[1] See Appendix II, page 116.

PROBLEMS

Solutions (or hints at solutions) to the problems marked with an asterisk (*) are given on page 123.

*1. Prove that every rational number q is *uniquely* expressible in the form $q = (b/c)p^n$, where b, c, and n are integers; c is positive; $(b, c) = 1$; and $p \nmid bc$. Here, as always in this section, p is any arbitrary but fixed prime. [HINT: See the remark on page 29.]

*2. (a) Show that $|360|_p \leq 1$ for every prime $p \neq 2, 3, 5$.

*(b) Show that $|360|_p = 1$ for every prime $p \neq 2, 3, 5$.

(c) Show that $|360|_p \cdot \prod_p |360|_p = 1$, where p runs through the set of all primes.

[HINT: For problem 2(a), (b), and (c), refer to the original Example 2.3.1 on pages 29–30.]

(d) Prove that $|b| \cdot \prod_p |b|_p = 1$ for every integer b.

(e) Prove finally that $|q| \cdot \prod_p |q|_p = 1$ for every nonzero rational number q. (This result is referred to as the **Product Formula** for the field of rational numbers Q.) [HINT: Use the result of part (d) in conjunction with the multiplicativity of the function $|\cdot|_p$.]

*3. In the spirit of the remark on page 30, determine the p-adic size of q for the following choices of p and q:

*(a) $p = 5$; $q = 105/1{,}000$, $q = 105$, $q = 1/105$, $q = 1{,}000$, $q = 1/1{,}000$.

*(b) $p = 2$; $q = 2^n$, where n runs through all positive integral values.

*(c) p is any arbitrary but fixed odd prime; $q = 2^n$, where n runs through all positive integral values.

4.(a) Is the function $|\cdot|_p$ **additive**—i.e., is it correct to claim that $|q + r|_p = |q|_p + |r|_p$ holds for every choice of the rational numbers q and r and for every choice of the prime p? Prove that your answer is correct.

*(b) Does the function $|\cdot|_p$ satisfy the standard **Triangle Inequality** (as does the standard absolute value function $|\cdot|$)—i.e., is it correct to claim that $|q + r|_p \leq |q|_p + |r|_p$ for every choice of the rational numbers q and r and for every choice of the prime p? Prove that your answer is correct. [HINT (strong!): Use the Ultrametric Inequality.]

*(c) Does the standard absolute value function $|\cdot|$ satisfy the Ultrametric Inequality? Prove that your answer is correct.

(d) Prove that $|q + r|_p = \max\{|q|_p, |r|_p\}$ whenever $|q|_p \neq |r|_p$.

*5. This problem presupposes some knowledge of sequences and series (see Appendix I, page 108). **Definition:** We say that the sequence $\{q_k\}_{k=1}^{+\infty}$ of rational numbers **converges *p*-adically** to L if and only if the sequence $\{|q_k - L|_p\}_{k=1}^{+\infty}$ of nonnegative real numbers converges (in the usual sense) to zero. Here, as usual, p is any arbitrary but fixed prime.

***(a)** Prove that the sequence $\{p^k\}_{k=1}^{+\infty}$ converges p-adically to zero.

(b) Prove that the sum of the infinite series $\sum_{i=0}^{\infty} 2^i$ is -1, provided that convergence is understood to be 2-adic convergence. **Remark:** I like to think that Euler, who was before his time in any case, was thinking 2-adically when he claimed that $-1 = 1 + 2 + 2^2 + \ldots + 2^i + \ldots$. [HINT: The sequence of partial sums has as a generic element a geometric progression that can (and should!) be summed.]

***(c)** Prove that $\sum_{n=1}^{\infty} n \cdot n!$ converges p-adically to -1, regardless of the choice of the prime p. [HINT: Use mathematical induction (see Appendix I, page 106) to get a closed form for $s_n = (1 \cdot 1!) + (2 \cdot 2!) + \ldots + (n \cdot n!)$ = the nth partial sum of the series. This closed form can be described in terms of the factorial function.]

†(d) Prove that the sequence $\{k!\}_{k=1}^{+\infty}$ converges p-adically to zero, regardless of the prime p.

6. The beginning of **ultrametric geometry.**

(a) Let p be any arbitrary but fixed prime, as usual. For every choice of the three rational numbers q, r, and u, we say that we have a **triangle** with vertices q, r, and u. We define the **length** of the **side** of the triangle joining vertices q and r as being equal to $|q - r|_p$—and likewise for the other two sides. With these definitions, prove that every triangle has at least two sides of equal length—i.e., that there are no scalene triangles! [HINT (strong!): Use the Ultrametric Inequality and, specifically, the result of problem 4(d) on page 32.]

(b) We define an **(open) sphere** of radius r and center q (r and q being rational numbers, as usual) to consist of all rational numbers x for which $|x - q|_p < r$, with p any arbitrary but fixed prime. Prove that every rational number c satisfying $|c - q|_p < r$ (i.e., belonging to the (open) sphere of radius r and center q) deserves to be called a center of the sphere, in the sense that the totality of rational numbers y satisfying $|y - c|_p < r$ coincides exactly with the totality of rational numbers x satisfying $|x - q|_p < r$. [HINT: Proceed as suggested in part (a) of this problem.]

7. Assume that b and $c > 0$ are integers and that $p^e \| (b^c - 1)$, where p is an odd prime and $e > 0$. Prove that $p^e \| c(b^{(c,p-1)} - 1)$. (See page 24, problem 13†.)

***8.*(a)** Does the function $|\cdot|^2$ satisfy the Triangle Inequality? Proof?

***(b)** Does the function $|\cdot|_p^2$ satisfy the Triangle Inequality? Proof?

*9. Assume that q is any arbitrary but fixed prime, and consider the infinite sequence $\{q^k\}_{k=1}^{+\infty}$.

*(a) What is the limit—if it exists—of this sequence relative to the usual absolute value?

*(b) What is the limit—if it exists—of this sequence relative to $|\cdot|_q$?

*(c) What is the limit—if it exists—of this sequence relative to $|\cdot|_p$, where p is any arbitrary but fixed prime other than q? Proof?

NOTES TO CHAPTER 2

1. The numbers f_n defined in problem 5 on page 22 are the Fermat numbers: $f_0 = 3$, $f_1 = 5$, $f_2 = 17$, $f_3 = 257$, and $f_4 = 65{,}537$—all of which are primes. Fermat conjectured that f_n is prime for all positive integers n, but Euler proved that $641|f_5$. At present no other prime Fermat numbers have been found, and the most popular current conjecture is that all Fermat numbers other than those Fermat himself showed to be prime are composite—the antithesis of Fermat's original conjecture!

2. Problem 10 on page 23 is designed to raise and answer some questions about how the quantities (b, c, d), $[b, c, d]$, and bcd are related. In Theorem 1.2.2 on page 11, we found that $(b, c)[b, c] = bc$ for all pairs of positive integers b and c. Problem 4(b) on page 13 told us that the naive guess that, for all triples b, c, and d of positive integers, $(b, c, d)[b, c, d] = bcd$ was wrong. Now parts (b) and (c) of problem 10 on page 23 provide us with "consolation prize" results. For further generalizations, see page 984 of the December 1961 issue of the *American Mathematical Monthly:* "Least Common Multiples and Highest Common Factors" by Marlow Sholander.

3. If the number $M_p = 2^p - 1$ is prime, it is referred to as a **Mersenne prime.** As parts (b) and (d) of problem 11 (page 23) point out, if M_p is to be prime it is necessary but (highly!) insufficient that p itself be a prime. As of October, 1987, there were precisely thirty Mersenne primes known; the largest of these is $M^{216,091}$, an integer that has in excess of 65,000 digits when expressed in the decimal scale and holds the record for being the largest known prime at present!

4. The results of problem 12 on page 24 are vestigial forms of a truly massive theorem of Dirichlet: If $(b, m) = 1$, then there are infinitely many primes in the arithmetic progression (see page 43, problem 6 for a definition of this term) $\{b, b + m, b + 2m, b + 3m, \ldots\}$. This famous result (circa 1839) is usually referred to as **Dirichlet's Theorem on Primes in Arithmetic Progression.** It is obvious that the condition $(b, m) = 1$ is necessary (why?), but all known proofs of the sufficiency of this same condition use advanced techniques.

5. Example 2.2.2 (pages 26–27) and the problems on pages 27–29 can be thought of as a glimpse into the world of **algebraic number theory.** Harry Pollard and Harold G. Diamond's *The Theory of Algebraic Numbers* remains one of the most readable introductions to this fascinating subject.
6. We have the genius of Kurt Hensel to thank for the invention of p-adic numbers. *Introduction to* p*-adic Numbers and Valuation Theory* by George Bachman is an extremely well-written book on the subject.

3 Numerical Congruences

INTRODUCTION

The concept permeating this chapter is that of **congruence**—not in the familiar geometrical sense of the word, but in a number-theoretic sense. We owe the idea of congruence, among many other things, to Gauss. It is undoubtedly one of the greatest notational breakthroughs in the entire history of mathematics. In some sense the definition of *congruence* is the definition of *divisibility* all over again; but from the point of view of abstract algebra, the congruence relation is a so-called **equivalence relation,** whereas the divisibility relation is not,[1] and this makes all the difference in the world. Having first introduced and become somewhat familiar with the congruence relation and some of its basic properties (and, also important, its "nonproperties"), we proceed to investigate a few of the many ramifications of this fundamental definition: **complete** and **reduced residue systems, linear** and **quadratic congruences,** the **order** symbol, and **primitive roots,** to name just a few. The **Euler ϕ-function,** an example of what are usually referred to as **arithmetic** or **number-theoretic functions** is introduced in Section 3.2 and studied in somewhat greater detail in Section 3.5. The major novel features of this chapter are, first, the systematic use of the p-adic approach introduced in Section 2.3 (see, specifically, the proof of the **Chinese Remainder Theorem** in Section 3.4) and, second, the strong emphasis on the order symbol, because of its great utility and importance.

Finally, Chapter 3 is where **abstract algebra** begins to make its presence felt in our treatment of elementary number theory. No specific background in abstract algebra is required, and I have tried my best to make the treatment totally self-contained. On the other hand, it would be foolish to deny that a student who has

[1]See Appendix I, page 106.

previously been exposed to abstract algebra—even if that exposure was superficial—has a built-in advantage at this point. **Appendix II (Abstract Algebra)** provides a brief introduction to the subject, with emphasis on the incredibly close connection between elementary number theory and elementary abstract algebra.

3.1 The Congruence Relation: Its Properties and "Nonproperties"

We begin with the basic definition of this chapter.

Definition 3.1.1: Let b, c, and $m > 1$ be integers. We say that b is **congruent** to c modulo m and write $b \equiv c \pmod{m}$ if and only if $m | (b - c)$. If $m \nmid (b - c)$, we write $b \not\equiv c \pmod{m}$ and say that b is not congruent (or is incongruent) to c modulo m.

Example 3.1.1: $13 \equiv 3 \pmod{5}$ because $5 | (13 - 3) = 10$, whereas $16 \not\equiv 24 \pmod{3}$ because $3 \nmid (16 - 24) = -8$. More generally, $b + jm \equiv c + km \pmod{m}$ for integers b, c, j, k, and $m > 1$ if and only if $m | (b + jm) - (c + km) = (b - c) + (j - k)m$. Now $m | (j - k)m$, regardless of the choice of the integers j and k, so $b + jm \equiv c + km \pmod{m}$ if and only if $m | (b - c)$—i.e., if and only if $b = c \pmod{m}$. This little fact was deduced by using the old idea of integral linear combinations (see part 4 of Theorem 1.1.1 on page 2).

Remark: It is extremely important to be able to pass back and forth from the congruence notation $b \equiv c \pmod{m}$ to the corresponding divisibility statement $m | (b - c)$ and from there to the corresponding equation, $b - c = km$ for some uniquely determined integer k.

Our next job is to start collecting some of the basic properties of the congruence notation. We make a package out of the first three properties because of the abstract algebraic interpretation alluded to in the chapter introduction.

Theorem 3.1.1: Assume that b, c, d, and $m > 1$ are integers.

1. $b \equiv b \pmod{m}$ for every integer b and for every integer $m > 1$. (This is referred to as the **reflexive** property of the congruence relation.)
2. If $b \equiv c \pmod{m}$, then $c \equiv b \pmod{m}$ (the **symmetric** property).

3. If $b \equiv c \pmod{m}$ and $c \equiv d \pmod{m}$, then $b \equiv d \pmod{m}$ (**transitivity**—compare part 2 of Theorem 1.1.1 on page 2).

Proof of part 3: Since we have only the definition of the congruence relation at present, our proof must be made on the basis of that definition. Let's systematically convert the two hypotheses and the desired conclusion into divisibility language via the definition and then see where we stand. The two hypotheses, $b \equiv c \pmod{m}$ and $c \equiv d \pmod{m}$, become $m|(b - c)$ and $m|(c - d)$. The desired conclusion becomes $m|(b - d)$. If $m|(b - c)$ and $m|(c - d)$ are assumed, however, then $m|(b - d)$ can easily be deduced because $b - d = (b - c) + (c - d)$; and now the integral linear combination property of the divisibility relation (see part 4 of Theorem 1.1.1 on page 2) allows us to conclude that $m|(b - d)$.

Remarks:

1. The practical effect of the symmetry property of the congruence relation is that you don't have to worry about whether you write $b \equiv c \pmod{m}$ or $c \equiv b \pmod{m}$ because either both statements are true or both statements are false.
2. You might wonder why it was assumed that m is greater than one in the definition of the congruence relation. This assumption is just a convenience—sort of a notational cleansing mechanism—rather than a necessity. We can't allow $m = 0$, mind you, because divisibility by zero was shown (see page 1) to be nonsense. We could allow m to be any negative integer, but this doesn't lead to any new information. To illustrate this point, consider the numerical fact $-5|4 - (-6) = 10$. If we were to extend the definition of the congruence relation to encompass negative integral values of m, then $-5|4 - (-6)$ could be expressed in the congruence notation as $4 \equiv -6 \pmod{-5}$. However, the mathematical content of the divisibility statement $-5|4 - (-6)$ is precisely the same as that of $+5|(-6) - 4$, which converts into the congruence statement $(-6) \equiv -4 \pmod{5}$; and now the modulus m is greater than one. Said otherwise, the assumption $m > 1$ involves no loss of generality—just as there is no loss of generality, when working with divisibility, in confining our attention to the positive integers (see remark 2 on page 2).

Now that we know that the congruence relation possesses the reflexive, symmetric, and transitive properties, we go on to find out how the congruence relation behaves relative to the so-called **rational operations** of addition, subtraction, multiplication, and division. These facts constitute the beginning of what could be referred to as "the algebra of the congruence notation" and are indispensable as basic tools for the manipulation of congruences.

Theorem 3.1.2: Assume that all symbols used are integers and that $m > 1$.

1. If $b \equiv c \pmod{m}$ and $d \equiv e \pmod{m}$, then $b + d \equiv c + e \pmod{m}$.
2. If $b \equiv c \pmod{m}$ and $d \equiv e \pmod{m}$, then $b - d \equiv c - e \pmod{m}$.
3. If $b \equiv c \pmod{m}$ and $d \equiv e \pmod{m}$, then $bd \equiv ce \pmod{m}$.
4. If $kb \equiv kc \pmod{m}$, then $b \equiv c \pmod{m/(k, m)}$.

Proof of Part 3: As in the proof of part 3 of Theorem 3.1.1 (and for the same reason), we immediately convert our congruence problem into the corresponding divisibility problem; i.e., if $m|(b - c)$ and $m|(d - e)$, then we need to conclude that $m|(bd - ce)$. The quantity bd appears in the statement of the conclusion we want, so at some point in our argument we have to get bd into the act. The quantity b appears in the assumption $m|(b - c)$, so we can claim that $m|d(b - c) = (bd - cd)$, using the appropriate property of the divisibility symbol; and now $m|(bd - cd)$ involves the quantity bd. Similarly, we deduce that $m|c(d - e) = cd - ce$ from the assumption $m|(d - e)$. We now have $m|(bd - cd)$ and $m|(cd - ce)$, and so $m|(bd - ce)$ (why?) as required.

Proof of Part 4: We are assuming that $m|(kb - kc) = k(b - c)$ or, equivalently, $m/(k, m)|k(b - c)/(k, m) = (k/(k, m))(b - c)$. (Give a reason justifying this last step.) Now $(m/(k, m), k/(k, m)) = 1$, so Theorem 1.2.1 on page 10 allows us to conclude that $m/(k, m)|(b - c)$. In congruence notation, $m/(k, m)|(b - c)$ is $b \equiv c \pmod{m/(k, m)}$, and so we are done.

Remarks

1. An informal way of expressing the first three parts of the theorem just proved is to say that congruences can be added, subtracted, and multiplied together. Please notice, however, that the moduli of the two input congruences don't get added, subtracted, or multiplied together! The modulus of the output congruence is the same as the modulus of each of the two input congruences.
2. Part 4 of the theorem just proved is by way of a warning that the procedure of division in a congruence isn't as straightforward as are addition, subtraction, and multiplication. For instance, it may be tempting (?) to conclude, on the basis of the assumption $kb \equiv kc \pmod{m}$, that $b \equiv c \pmod{m}$, but this conclusion cannot in general be drawn (proof?). We may regard the valid conclusion $b \equiv c \pmod{m/(k, m)}$ as a sort of consolation prize, in place of the hoped-for conclusion $b \equiv c \pmod{m}$.

Example 3.1.2: $60 \equiv 90 \pmod{10}$ and $12 = 60/5 \equiv 18 = 90/5 \pmod{2 = 10/5}$, but notice that $12 \not\equiv 18 \pmod{10}$.

We now establish a result, known as Wilson's Theorem, that gives us another way of deciding whether or not a positive integer is a prime. Problem 12 on page 44 makes the foregoing rather loose statement precise.

Wilson's Theorem: If p is a prime, then $(p - 1)! \equiv -1 \pmod{p}$.

Before undertaking a proof, let's look at the technique we will employ in the proof in a numerical example.

Example 3.1.3: If $p = 13$, we may pair off the $p - 1 = 12$ factors of $(p - 1)! = (13 - 1)! = 12!$ as follows: $1 \leftrightarrow 12$, $2 \leftrightarrow 7$, $3 \leftrightarrow 9$, $4 \leftrightarrow 10$, $5 \leftrightarrow 8$ and $6 \leftrightarrow 11$. Here the notation $b \leftrightarrow c$ means that b corresponds to c in the pairing (see also page 5, problem 4). Notice that $1 \cdot 12 \equiv -1 \pmod{13}$, while the product of each of the remaining five pairs is congruent to $+1$ modulo 13; i.e., $2 \cdot 7 \equiv 1 \pmod{13}$, $3 \cdot 9 \equiv 1 \pmod{13}$, $4 \cdot 10 \equiv 1 \pmod{13}$, $5 \cdot 8 \equiv 1 \pmod{13}$, and $6 \cdot 11 \equiv 1 \pmod{13}$. Thus $12! = (1 \cdot 12)(2 \cdot 7)(3 \cdot 9)(4 \cdot 10)(5 \cdot 8)(6 \cdot 11) \equiv (-1)(+1)(+1)(+1)(+1)(+1) \equiv -1 \pmod{13}$, and Wilson's Theorem has been verified for the prime $p = 13$.

Proof of Wilson's Theorem: $(2 - 1)! \equiv -1 \pmod{2}$, so we may assume now that p is an odd prime. Hence $(p - 1)!$ is an even positive integer. The idea of the proof is to take the $p - 1$ positive integers $1, 2, 3, \ldots, (p - 1)$, which are the individual factors of $(p - 1)!$, and pair these factors off in a special way—as illustrated in the foregoing example—so as to prove the result. We now describe the mechanics of this pairing process for any arbitrary but fixed odd prime p. The set $\{1, 2, 3, \ldots, p - 1\}$ consists of $(p - 1)/2$ pairs (i, j), where we may assume that $1 \leqslant i < j \leqslant p - 1$ without loss of generality. We start by pairing 1 with $p - 1$, so that $1 \leftrightarrow p - 1$. Notice that the product, $1 \cdot (p - 1)$, of this particular pair is congruent to -1 modulo p. Now preassign i subject to $2 \leqslant i \leqslant (p - 2)$. Our job is to prove the existence and uniqueness of the integer j satisfying $i < j \leqslant p - 2$ and $ij \equiv +1 \pmod{p}$. First, the congruence $iX \equiv 1 \pmod{p}$ has a solution, say x, because $ix \equiv 1 \pmod{p}$ is equivalent to $p|(ix - 1)$, which in turn is equivalent to $ix - 1 = (-y)p$ for some integer y, or finally to $ix + py = 1$; and this equation has solutions in integers x and y because $(i, p) = 1$ (see page 9, remark 3). Next, if x_1 and x_2 are any two solutions of the linear (in X) congruence $iX \equiv 1 \pmod{p}$, then $ix_1 \equiv ix_2 \equiv 1 \pmod{p}$; from this it follows that $ix_1 \equiv ix_2 \pmod{p}$ and hence that $x_1 \equiv x_2 \pmod{p}$—upon referring back to page 40, part 4 of Theorem 3.1.2. This last fact allows us to conclude that the linear (in X) congruence $iX \equiv 1 \pmod{p}$ has precisely one solution j satisfying $0 \leqslant j \leqslant p$ (why?). I will leave it to the reader (see page 44, problem 11) to check that $j \neq 0$, 1, or $(p - 1)$, so that $2 \leqslant j \leqslant (p - 2)$. If $i \geqslant j$, we simply interchange the labeling so that $i \leqslant j$. We need to prove that $i \neq j$, so that we have a pair of distinct elements of the set $\{2, 3, 4, \ldots, (p-2)\}$. If $i = j$, then $ij \equiv 1 \pmod{p}$ becomes $i^2 \equiv 1 \pmod{p}$, which converts into $p|(i^2 - 1) = (i+1)(i-1)$, from which it would follow that $p|(i+1)$ or $p|(i-1)$. However, $1 \leqslant i - 1 < i + 1 \leqslant p - 1$ because of the assumptions on i, so $p \nmid (i+1)$ and $p \nmid (i-1)$—a contradiction. Hence, $i \neq j$. Now $(p-1)! \equiv 1 \cdot (p-1) \cdot \Pi\, ij$, where the product runs over the $(p-1)/2$ pairs (i, j). Looking at this equation modulo p, we get

$$(p-1)! \equiv 1 \cdot (p-1) \cdot \Pi\, ij \equiv 1 \cdot (p-1) \cdot 1^{(p-1)/2} \equiv -1 \pmod{p}$$

because $p - 1 \equiv -1 \pmod{p}$, and every (i, j) pair was built in such a way so as to have $ij \equiv +1 \pmod{p}$.

Remarks:

1. The proof of Wilson's Theorem just presented is fairly long—primarily because we don't know that much about the congruence relation yet. We shall be able to get shorter proofs later on in the text, after the requisite theory has been developed (see Example 4.1.2, page 85). On the other hand, this first proof has the advantage of being rather "organic": it is based entirely on one simple but important idea—namely, that the factors entering $(p-1)$ can be paired off.
2. In the course of the proof, we learned something about solving some rather specialized types of linear congruences. In Section 3.3 of this chapter, we will study linear and quadratic congruences in somewhat greater generality.

PROBLEMS

Solutions (or hints at solutions) to the problems marked with an asterisk (*) are given on page 124.

*1. The integer b is said to be a solution of the **linear** (in X) **congruence** $X \equiv c \pmod{m}$ if and only if $b \equiv c \pmod{m}$. Find all solutions of each of the following linear congruences that also satisfy the condition $|b| < 50$:
   (a) $X \equiv 2 \pmod{8}$
   (b) $X \equiv 13 \pmod{19}$
   (c) $2X \equiv 1 \pmod{5}$
   (d) $X \equiv -14 \pmod{8}$
   ***(e)** $9X \equiv 12 \pmod{15}$
   ***(f)** $9X \equiv 13 \pmod{15}$
2. Prove that $b \equiv r \pmod{m}$ if and only if b leaves a remainder of r upon division by m, in the sense of the Division Algorithm (see page 3). Find all integers b satisfying $|b| < 50$ that leave a remainder of 5 upon division by 12.

*3. (a) Prove that the square of any integer, when expressed in the decimal scale, is a number ending in 0, 1, 4, 5, 6, or 9; i.e., the so-called *units digit* (the digit appearing at the extreme right-hand end in the decimal expansion of the number in question) is always 0, 1, 4, 5, 6, or 9.
(b) What is the result analogous to the result in part (a) for fourth powers? Prove that your answer is correct.
***(c)** Prove that the cube of any integer leaves a remainder of 0, 1, or 8 upon division by 9.

4. (a) Prove that the congruence relation is **reflexive**—i.e., $b \equiv b \pmod{m}$ for every integer b and for every integer $m > 1$.
(b) Prove that the congruence relation is **symmetric**—i.e., if $b \equiv c$ (mod m), then $c \equiv b \pmod{m}$.

*5. (a) Prove that congruences can be added and subtracted in the sense of parts 1 and 2 of Theorem 3.1.2 on page 40.
***(b)** Is it correct to claim that, if $b \equiv c \pmod{m}$ and if $r \equiv s \pmod{m}$, then $bx + ry \equiv cx + sy \pmod{m}$, regardless of the choice of the integers x and y? Prove that your answer is correct.
(c) How are parts (a) and (b) of this problem related? Be as precise as you can.

*6. The **arithmetic progression** containing the integer b and having common difference m is the set $\{\ldots, (b - 2m), (b - m), b, (b + m), (b + 2m), \ldots\}$. In parts (a) and (b) below, find all real numbers x satisfying $|x| < 50$ that belong to the following arithmetic progressions.
(a) The arithmetic progression containing the integer 7 and having common difference 10.
(b) The arithmetic progression containing the integer -4 and having common difference -5.
***(c)** In congruence notation, express the conditions imposed on the integers x in parts (a) and (b).

*7. In each of the following, decide whether the statement is correct, and then justify your answer; i.e., if your answer is yes, provide a proof, and if your answer is no, provide a counterexample.
***(a)** $b \equiv c \pmod{m}$ implies that $b \equiv c \pmod{d}$ for every integer $d > 1$ satisfying $d|m$.
***(b)** If $d|m$ and $b \equiv c \pmod{d}$, then $b \equiv c \pmod{m}$.
***(c)** If $b \equiv c \pmod{m/(k, m)}$, then $kb = kc \pmod{m}$. [NOTE: This is the converse of part 4 of Theorem 3.1.2 on page 40.]
***(d)** If $b \equiv c \pmod{m/(k, m)}$, then $kb \equiv kc \pmod{km/(k, m)}$.
***(e)** If $b = c$, then $b \equiv c \pmod{m}$ for every integer $m > 1$.
***(f)** If $b \equiv c \pmod{m}$ for every integer $m > 1$, then $b = c$.
***(g)** If $b \equiv c \pmod{m/(k, m)}$, then $(k, m)b \equiv (k, m)c \pmod{m/(k, m)}$.
***(h)** If $kb \equiv kc \pmod{m}$ and k and m are relatively prime, then $b \equiv c \pmod{m}$.

***8.** (a) Prove that, if $b \equiv c \pmod{m_1}$ and $b \equiv c \pmod{m_2}$, then $b \equiv c \pmod{[m_1, m_2]}$, where $[m_1, m_2]$ is the least common multiple (LCM) of m_1 and m_2 (see page 11).

***(b)** Find a single linear congruence of the form $X \equiv b \pmod{m}$ for suitable choices of b and $m > 1$ that has precisely the same solution set as does the pair of simultaneous linear congruences, $X \equiv 1 \pmod{3}$ and $X \equiv 2 \pmod{4}$; i.e., you need to discover values of b and $m > 1$ so that the following condition is true: the integer z satisfies $z \equiv b \pmod{m}$ if and only if $z \equiv 1 \pmod{3}$ and $z \equiv 2 \pmod{4}$. [HINT: Reinterpret each of the linear congruences $X \equiv 1 \pmod{3}$, $X \equiv 2 \pmod{4}$ in terms of the new modulus $[3, 4]$; i.e., $X \equiv 1 \pmod{3}$ if and only if $X \equiv\ ? \pmod{[3, 4]}$.]

***9.** Prove that every integer x satisfies at least one of the following linear congruences: $X \equiv 0 \pmod{2}$, $X \equiv 0 \pmod{3}$, $X \equiv 1 \pmod{4}$, $X \equiv 5 \pmod{6}$, $X \equiv 7 \pmod{12}$. Are there any integers which satisfy two or more of these congruences? [HINT: Work modulo $[2, 3, 4, 6, 12]$. Every integer x satisfies $x \equiv b \pmod{12}$ for precisely one value of b coming from the set $\{0, 1, 2, 3, 4, 5, 6, 7, 8, 9, 10, 11\}$.]

***10.** Assume that $f(X) = b_nX^n + b_{n-1}X^{n-1} + \ldots + b_1X^1 + b_0$ is a polynomial in (the indeterminate) X that has integral coefficients $b_n, b_{n-1}, \ldots, b_1, b_0$. Assume further that $b \equiv c \pmod{m}$, and prove that $f(b) \equiv f(c) \pmod{m}$. [HINT (strong!): Congruences can be added, subtracted, and multiplied together, according to parts 1, 2, and 3 of Theorem 3.1.2 on page 40. The polynomial involves only these three rational operations.]

11. Verify that $j \neq 0$, 1, or $p - 1$, as suggested on page 41 in the proof of Wilson's Theorem.

***12. Converse of Wilson's Theorem.** Prove that $(m - 1)! \not\equiv -1 \pmod{m}$ if m isn't prime. [HINT: Do this by "brute force" (i.e., number-crunching) for $m = 4$ and do it for every $m > 4$ by establishing that $(m - 1)! \equiv 0 \pmod{m}$.]

***13.*(a)** If $b \equiv c \pmod{m}$ and $d \equiv e \pmod{m}$, does it always follow that $bd \equiv ce \pmod{m}$? Prove that your answer is correct.

***(b)** If $m \equiv 0 \pmod{p}$ for infinitely many distinct primes p, what can you conclude about the integer m?

***14.*(a)** Prove that the congruence relation is reflexive, symmetric, and transitive, using the technique of p-adic valuations.

(b) Prove that $kb \equiv kc \pmod{m}$ if and only if $b \equiv c \pmod{m/(k, m)}$, using the technique of p-adic valuations.

***15.** Prove that, if $r = (p - 1)/2$, then $(r!)^2 \cdot (-1)^r \equiv -1 \pmod{p}$ whenever p is an odd prime. [HINT: Use the same sort of technique as was used in the proof of Wilson's Theorem, page 41.]

*16. If p is a prime and if $h + k = p - 1$ for nonnegative integers h and k, prove that $h!k! + (-1)^h \equiv 0 \pmod{p}$.

17. Prove that $b \equiv c \pmod{m}$ is equivalent to the requirement that $|(b - c)/m|_p \leqslant 1$ for every prime p. Now use this p-adic criterion to create your own proof of the fact (see Theorem 3.1.1, page 38) that the congruence relation is an equivalence relation.

*18. A rational number q is integral if and only if $|q|_p \leqslant 1$ for every prime p; this is a p-adic criterion for "integralness." Apply this criterion to produce proofs of parts 1 and 4 of Theorem 3.1.2 on page 40.

3.2 Complete Residue Systems, Reduced Residue Systems, and an Introduction to the Euler ϕ-Function

We begin by introducing the two closely related and important concepts of **complete residue system** and **reduced residue system**. Suppose that we have preassigned the modulus $m = 6$. Consider the set $\{0, -5, 17, 10, 3, -10\}$, and notice that it has the following properties:

1. The elements of our set are pairwise incongruent modulo m. For instance, $-5 \not\equiv 17 \pmod{6}$ because $6 \nmid (-5 - 17) = -22$, etc.
2. Regardless of the choice of the integer z, z is congruent (mod $6 = m$) to at least one (and hence, exactly one—why?) element of our set.

We call the set $\{0, -5, 17, 10, 3, -10\}$ a complete residue system modulo 6, and we make the following definition.

Definition 3.2.1: Assume that m is a positive integer. Every set C of integers satisfying the two conditions—first, that $x \not\equiv y \pmod{m}$ for every pair of distinct elements of C, and second, that every integer is congruent modulo m to some element of C—is said to be a **complete residue system modulo *m*.**

Remarks: It follows from the above definition that every complete residue system modulo m, call it C, has two further properties—namely:

3. C contains m elements.
4. Every integer is congruent modulo m to precisely one element of C.

The proofs of these two facts will be left as problems (see problem 4 on page 51).

Example 3.2.1: The two most popular complete residue systems modulo m are $\{0, 1, 2, \ldots, m - 1\}$ and $\{1, 2, 3, \ldots, m\}$. If m is odd, another choice of complete residue system modulo m that sometimes proves useful is $\{0, \pm 1, \pm 2, \ldots, \pm (m - 1)/2\}$. Please check that each of the first two sets is a complete residue system modulo m and that the third is a complete residue system modulo m whenever the modulus m is odd.

Example 3.2.2: The complete residue system modulo 12 consisting of the elements of the set $\{0, 1, 2, 3, 4, 5, 6, 7, 8, 9, 10, 11\}$ can be regarded as having been constructed from the five linear congruences $X \equiv 0 \pmod 2$, $X \equiv 0 \pmod 3$, $X \equiv 1 \pmod 4$, $X \equiv 5 \pmod 6$, and $X \equiv 7 \pmod{12}$, as follows: each of the integers 0, 2, 4, 6, 8, and 10 satisfies $X \equiv 0 \pmod 2$, and no two of these integers are congruent modulo 12; so we start building C, our complete residue system modulo 12, by requiring that it contain the set $\{0, 2, 4, 6, 8, 10\}$ as a (proper) subset. The single linear congruence $X \equiv 0 \pmod 3$ is equivalent to the $4 = 12/3$ linear congruences $X \equiv 0 \pmod{12}$, $X \equiv 3 \pmod{12}$, $X \equiv 6 \pmod{12}$, and $X \equiv 9 \pmod{12}$. The integers 0 and 6 have already been required to belong to our complete residue system modulo 12, but now we further require that 3 and 9 belong to C; so now, in total, we have $\{0, 2, 3, 4, 6, 8, 9, 10\}$ contained in C. Note that the eight elements of the set, 0, 2, 3, 4, 6, 8, 9, and 10, are still pairwise incongruent modulo 12. To fill out C, we require precisely four more integers (why?) to be chosen in such a way that no two elements of the final set C are congruent modulo 12. Continuing in the same way as we have in our example so far, we can use the congruence $X \equiv 1 \pmod 4$ to contribute 1 and 5 to C, $X \equiv 5 \pmod 6$ to contribute the number 11 to C, and finally $X \equiv 7 \pmod{12}$ to contribute the number 7 to C. Thus $C = \{0, 1, 2, 3, 4, 5, 6, 7, 8, 9, 10, 11\}$ is the resulting complete residue system modulo 12. Now look back at problem 9 on page 44.

The idea of a reduced residue system modulo m is extremely similar to the idea of a complete residue system modulo m; the only difference is that our attention is now focused exclusively on integers that are relatively prime to the modulus in question.

Definition 3.2.2: Let $m > 1$ be an integer. Every set R of integers satisfying the two conditions—first, that $x \not\equiv y \pmod m$ for every pair of distinct elements of R, and second, that every integer relatively prime to m is congruent modulo m to some element of R—is said to be a **reduced residue system modulo *m*.**

Example 3.2.3: The set $\{1, 3, 5, 7\}$ is a reduced residue system modulo 8, while each of the sets $\{1, -1, 2, -2, 3, -3\}$ and $\{5, 25, 125, 625, 3{,}125, 15{,}625\}$ is a reduced residue system modulo 7. Please verify!

Remarks:

1. The term *reduced residue system modulo* m is sensible because every such set R can be thought of as coming from some complete residue system modulo m, say C, precisely through our deleting the elements of C that aren't relatively prime to the modulus m. For instance, we can obtain $\{1, 3, 5, 7\}$ from the complete residue system $\{0, 1, 2, 3, 4, 5, 6, 7\}$ with respect to the same modulus 8 by deleting the elements 0, 2, 4, and 6—precisely the elements that have some nontrivial factor in common with the modulus 8. See page 51, problem 6 for further examples.
2. It is, in general, trickier to produce a reduced residue system modulo m than to produce a complete residue system modulo m because of the constraint that all the elements of every reduced residue system modulo m must be relatively prime to the modulus m. To illustrate, $\{1, 2, 3, \ldots, m - 1\}$ is a reduced residue system modulo m if and only if m is a prime (why?).
3. It follows from the above definition that every reduced residue system modulo m has an additional property—namely, that every integer relatively prime to m is congruent modulo m to precisely one element of the given reduced residue system modulo m (why?).

We noted, in the remarks (page 45) following Definition 3.2.1, that every complete residue system modulo m must contain exactly m elements. It is certainly natural to ask whether every reduced residue system modulo m must contain the same number of elements and, if so, what that common number of elements might be.

Let us start the investigation by noticing that the set $\{r_1, r_2, \ldots, r_{\phi(m)}\}$, consisting precisely of all the positive integers r satisfying $1 \leqslant r \leqslant m$ and $(r, m) = 1$ constitutes a reduced residue system modulo m. Here the symbol $\phi(m)$ can be interpreted as a count of the total number of positive integers that are both less than or equal to m and relatively prime to m. First, $r_i \not\equiv r_j \pmod{m}$ is satisfied for every pair of distinct elements (r_i, r_j) of our set $\{r_1, r_2, \ldots, r_{\phi(m)}\}$ because, if $r_i < r_j$, then $0 < r_j - r_i \leqslant m - 1 < m$, and m can't be a divisor of an integer, such as $r_j - r_i$, which is strictly between 0 and m. [NOTE: Since r_i and r_j were assumed to be distinct elements of the original set, there is no loss of generality in assuming that $r_i < r_j$.] Secondly, every integer z satisfying $(z, m) = 1$ can be written (uniquely, as a matter of fact) in the form $z = qm + r$ where $1 \leqslant r < m$ (why can't r be zero?). If $(r, m) \neq 1$, then $(z, m) \neq 1$ (why?), contradicting our assumption that $(z, m) = 1$. So $(r, m) = 1$ and $1 \leqslant r \leqslant m$; and so $r = r_i$ for some suitable value of i, $1 \leqslant i \leqslant \phi(m)$; and so finally $z = qm + r \equiv r_i$

(mod m), as required by the definition of reduced residue system modulo m. To finish our job of showing that every reduced residue system modulo m has the same number of elements, it suffices to argue that every reduced residue system-modulo m has exactly $\phi(m)$ elements. I claim (you prove, please—see page 52, problem 8) that, if R is a set with fewer than $\phi(m)$ elements, then there must be some integer z, relatively prime to m, that is congruent to no element of R—which means that R can't be a reduced residue system modulo m. On the other hand, if R has more than $\phi(m)$ elements, then either $(r, m) > 1$ for some element of R or $r \equiv s \pmod{m}$ for some suitable distinct elements r and s of R—and so, once again, R can't be a reduced residue system modulo m.

From the preceding discussion, we make a definition.

Definition 3.2.3: The Euler ϕ-function (sometimes called the **Euler totient function**), ϕ, is defined by $\phi(m)$ = the number of positive integers r satisfying $1 \leqslant r \leqslant m$ and $(r, m) = 1$.

This arithmetic function has a long and honorable history. A great deal is known about the Euler ϕ-function, but it is nonetheless quite safe to claim that a great deal more is not known. We shall conclude this section by deriving some of the most immediate results concerning ϕ. The problems at the end of this section and, more importantly, Section 3.5 and the problems attached to that section are designed to help us to probe a little more deeply into the properties of the ϕ-function.

Both complete residue systems modulo m and reduced residue systems modulo m have, with respect to multiplication, a certain property of "stability" or "robustness" about them that is extremely useful. A precise formulation is as follows.

Theorem 3.2.1: Suppose that $\{z_1, z_2, \ldots, z_n\}$ is a complete or reduced residue system modulo m. Assume further that the integer b is relatively prime to the modulus m. Then the set $\{bz_1, bz_2, \ldots, bz_n\}$ is again a complete or, respectively, reduced residue system modulo m.

Proof: I'll do the case of the reduced residue system modulo m and ask you to do the case of the complete residue system modulo m (see page 53, problem 11). Reverting to our usual notation, let $\{r_1, r_2, \ldots, r_{\phi(m)}\}$ be the given reduced residue system modulo m. I need to prove that the set $\{br_1, br_2, \ldots, br_{\phi(m)}\}$ is, once again, a reduced residue system modulo m. Suppose that $br_i \equiv br_j$ (mod m) for some pair of subscripts (i, j) satisfying $1 \leqslant i < j \leqslant \phi(m)$. Then $m|b(r_i - r_j)$ and, since $(b, m) = 1$ by assumption, $m|(r_i - r_j)$; i.e., $r_i \equiv r_j \pmod{m}$.

However, this is impossible because the original set $\{r_1, r_2, \ldots, r_{\phi(m)}\}$ was a reduced residue system modulo m. Hence the set $\{br_1, br_2, \ldots, br_{\phi(m)}\}$ satisfies the first condition of Definition 3.2.2 (page 46). To prove that the set $\{br_1, br_2, \ldots, br_{\phi(m)}\}$ satisfies the second condition of Definition 3.2.2, it suffices to prove that, regardless of the choice of the integer z, there exists some subscript i, $1 \leqslant i \leqslant \phi(m)$, for which $z \equiv br_i \pmod m$. Since $(b, m) = 1$, there exists an integer, call it c, such that $bc \equiv 1 \pmod m$. (Recall the proof of Wilson's Theorem on page 41.) The statement $z \equiv br_i \pmod m$ is thus equivalent to the statement $cz \equiv (bc)r_i \pmod m$ or, ultimately, $cz \equiv r_i \pmod m$, using the fact that $bc \equiv 1 \pmod m$. However the existence of i for which $cz \equiv r_i \pmod m$ is guaranteed by virtue of the fact that $\{r_1, r_2, \ldots, r_{\phi(m)}\}$ was assumed to be a reduced residue system modulo m. The proof is now complete (why?).

Remark: We could have defined the concept of reduced residue system via a slightly altered version of Definition 3.2.2.

Definition 3.2.2′: Let $m > 1$ be an integer. Every set R of integers satisfying the two conditions—first, that $x \not\equiv y \pmod m$ for every pair of distinct elements of R, and second, that R has $\phi(m)$ elements—is said to be a **reduced residue system modulo m.**

We immediately put this "stability" property of reduced residue systems modulo m to work in order to obtain the following extremely useful result.

Theorem 3.2.2 – The Euler-Fermat Theorem: If $(b, m) = 1$, then $b^{\phi(m)} \equiv 1 \pmod m$.

Proof: Let $R = \{r_1, r_2, \ldots, r_{\phi(m)}\}$ be any arbitrary but fixed reduced residue system modulo m. Then because $(b, m) = 1$, $bR = \{br_1, br_2, \ldots, br_{\phi(m)}\}$ is again a reduced residue system modulo m. It follows that $\prod_{i=1}^{\phi(m)} r_i \equiv \prod_{i=1}^{\phi(m)} (br_i) \pmod m$ (why?), and $\prod_{i=1}^{\phi(m)} (br_i) = b^{\phi(m)} \prod_{i=1}^{\phi(m)} r_i$, because $\phi(m)$ is the number of elements of R. By the transitive property of the congruence relation, it now follows that $b^{\phi(m)} \prod_{i=1}^{\phi(m)} r_i \equiv \prod_{i=1}^{\phi(m)} r_i \pmod m$. Since $(r_i, m) = 1$ for each i such that $1 \leqslant i \leqslant \phi(m)$, it follows that $\left(\prod_{i=1}^{\phi(m)} r_i, m\right) = 1$ (page 10, property 5 of the GCD), and hence we may conclude (Theorem 3.1.2, part 4 on page 40) that $b^{\phi(m)} \equiv 1 \pmod{m}$.

The most important special case of the Euler-Fermat Theorem is expressed in the following corollary.

Corollary 3.2.1 – Fermat's Little Theorem: If p is a prime and if $p \nmid b$, then $b^{p-1} \equiv 1 \pmod{p}$.

Proof: The assumption that $p \nmid b$ is equivalent to the assumption that $(b, p) = 1$, because p is assumed here to be a prime. The exponent $\phi(m)$ in the Euler-Fermat Theorem becomes $\phi(p)$, and $\phi(p) = p - 1$ because $\{1, 2, \ldots, p - 1\}$ is the set of positive integers less than or equal to p and relatively prime to p.

The last result of this section (also a quick consequence of the Euler-Fermat theorem) tells us not only that the linear (in X) congruence $bX \equiv c \pmod{m}$ is solvable when $(b, m) = 1$ but also exhibits a specific solution of the congruence.

Corollary 3.2.2: If $m > 1$ is an integer and if $(b, m) = 1$, then the linear (in X) congruence $bX \equiv c \pmod{m}$ is solvable and has $x = b^{\phi(m)-1}c$ as a solution.

Proof: Obvious.

Linear and quadratic congruences are the subject of the next section, where we will be concerned not only with the solvability of these congruences but also with the number of solutions involved.

PROBLEMS

Solutions (or hints at solutions) to the problems marked with an asterisk (*) are given on page 125.

*1. Verify each of the following statements:
   (a) $\{0, 1, 2, \ldots, m - 1\}$ and $\{1, 2, 3, \ldots, m\}$ are always complete residue systems modulo m, regardless of the choice of the integer $m > 1$.
   *(b) $\{5, 25, 125, 625, 3{,}125, 15{,}625\}$ is a reduced residue system modulo 7.
   (c) $\{0, \pm 1, \pm 2, \ldots, \pm(m - 1)/2\}$ is a complete residue system modulo m if and only if m is odd.

*2. (a) Construct a complete residue system modulo 10 that consists entirely of integral multiples of 7.

***(b)** Construct a reduced residue system modulo 7 that consists entirely of integral powers of 3.

***3.** Assume that $R = \{r_1, r_2, \ldots, r_{\phi(m)}\}$ is the particular reduced residue system modulo m that consists of all positive integers less than or equal to m and relatively prime to m.

(a) For $m = 1, 2, 3, \ldots, 13$, exhibit this particular reduced residue system modulo m. Use the preceding to evaluate $\phi(1), \phi(2), \ldots, \phi(13)$.

***(b)** Is it always correct to claim—for this particular reduced residue system modulo m—that $r = 1$ and $r_{\phi(m)} = m - 1$? Why? [HINT: Consider
$m = 1$ and $m > 1$.]

***(c)** Prove that $\phi(m)$ is even, except in the initial "pathological" cases $m = 1$ and $m = 2$. [HINTS (strong): (1) Prove that r is an element of R if and only if $m - r$ is an element of R. (2) Is it possible to have $r = m - r$ if r is an element of R?]

***(d)** Prove that $r_1 + r_2 + \ldots + r_{\phi(m)} = \sum_{i=1}^{\phi(m)} r_i = (m \cdot \phi(m))/2$. [HINT: Build on what you have learned from doing part (c) of this problem.]

4. Assume that C is an arbitrary complete residue system modulo m. Prove that C must have the following properties:

(a) C contains m elements.

(b) Every integer is congruent modulo m to precisely one element of C.

***5.** Assume, again, that C is an arbitrary complete residue system modulo m.

***(a)** Could two (or more) elements of C belong to some arithmetic progression having common difference m? Prove that your answer is correct. (Refer to page 43, problem 6, for the definitions of *arithmetic progression* and *common difference*, if necessary.)

***(b)** Suppose $x_i = q_i m + r_i$, $0 \leqslant i < m$, upon applying the Division Algorithm to x_i for each element x_i of C. Could it happen that $r_i = r_j$ even though $i \neq j$? Prove that your answer is correct.

***6.** In each numerical example, delete the appropriate elements of the given complete residue system modulo m in order to produce the associated reduced residue system modulo m:

(a) $\{0, 1, 2, 3, 4, 5, 6, 7, 8, 9, 10, 11\}$

(b) $\{-2, -1, 0, 1, 2\}$

(c) $\{0, -5, 17, 10, 3, -10\}$

***(d)** $\{1, 2, 3, 4, 5, 6, 7, 8\}$

***7.** Assume that $R = \{r_1, r_2, \ldots, r_{\phi(m)}\}$ is an arbitrary reduced residue system modulo m.

***(a)** Could there exist some integer z satisfying $(z, m) = 1$ such that z is congruent modulo m to at least two elements of R? Prove that your answer is correct.

*(b) Could there exist some integer z such that z is congruent modulo m to at least two elements of $\{r_1, r_2, \ldots, r_{\phi(m)}\}$? Prove that your answer is correct.

*(c) Could there exist some integer z satisfying $(z, m) > 1$ such that $z \equiv r$ (mod m) for at least one value of i, $1 \leqslant i < \phi(m)$? Prove that your answer is correct.

8. If S is a set of integers containing fewer than $\phi(m)$ elements, then prove that there must exist some integer z satisfying $(z, m) = 1$ while $z \not\equiv s(\text{mod } m)$, regardless of the choice of s in S. (Hence, S cannot be a reduced residue system modulo m—look back at the argument on page 48.)

*9. (This problem requires some knowledge of the **Fibonacci sequence**—refer to page 6, problem 9.) We define the Fibonacci sequence $\{F_k\}_{k=1}^{+\infty}$ as follows:

$$F_0 = 1, F_1 = 1, F_k = F_{k-1} + F_{k-2}$$

for all integers $k \geqslant 2$.

(a) Compute $F_2, F_3, \ldots, F_{13}$.

(b) Compute $\phi(F_2), \phi(F_3), \ldots, \phi(F_{13})$.

*(c)†Is it correct to claim that $4|\phi(F_k)$ for all integers $k > 5$? Prove that your answer is correct.

*10. **Pythagorean triples.** We define the ordered triple of integers (x, y, z) to be a Pythagorean triple if and only if $x^2 + y^2 = z^2$. If $xyz = 0$, there are only trivial solutions of the types $(x, 0, \pm x)$, $(0, y, \pm y)$, so let us assume $xyz > 0$. In fact, we lose no important information by assuming $x > 0$, $y > 0$, and $z > 0$. Suppose that at least two of the positive integers x, y, z are divisible by the same prime p. Prove that this forces the third element of the Pythagorean triple to be divisible by p. This little fact tells us that, for instance, if $x = pX$ and $y = pY$ for appropriate positive integers X and Y, then there must exist some suitable positive integer Z for which $z = pZ$. But now (X, Y, Z) is also a Pythagorean triple (why?). The upshot of all this is that, again, we lose no information by assuming that x, y, and z are *pairwise relatively prime*—i.e., that $(x, y) = (x, z) = (y, z) = 1$ (compare problem 3(c) on page 13). For example, once we know the Pythagorean triple (3, 4, 5), we can get the other triples (6, 8, 10), (9, 12, 15), ... A Pythagorean triple (x, y, z) satisfying the further condition $(x, y) = (y, z) = (z, x) = 1$ is said to be a **primitive Pythagorean triple.**

(a) Prove that at least one of the two positive integers x and y is even if (x, y, z) is any Pythagorean triple. [HINT: $x^2 + y^2 = z^2$ requires that $x^2 + y^2 \equiv z^2$ (mod 4). What possibilities are there for z^2 modulo 4?]

(b) If (x, y, z) is any Pythagorean triple, then prove that at least one of the integers x and y is divisible by 3. [HINT: Work modulo 3.]

***(c)** Is it correct to claim that <u>at least one</u> of x, y and z is divisible by 4 if (x, y, z) is a Pythagorean triple? Prove that your answer is correct. [HINT: If w is an integer and $w^2 \equiv r \pmod 4$, what values can r take on? There is no loss of generality in assuming $0 \leqslant r \leqslant 3$.]

***(d)** Is it correct to claim that <u>at least one</u> of x, y, and z is divisible by 5 if (x, y, z) is a Pythagorean triple? Prove that your answer is correct. [HINT: If w is an integer, and if $w^2 \equiv r \pmod 5$ and if, without loss of generality, $0 \leqslant r \leqslant 4$, then what values are available to r?]

***(e)** Does there exist an integer $n > 5$ such that <u>at least one</u> of x, y, and z is divisible by n, regardless of the choice of the Pythagorean triple (x, y, z)? Prove that your answer is correct.

***(f)** Must <u>at least one</u> of x, y, and z be prime in every primitive Pythagorean triple, as is the case in the examples (3, 4, 5), (5, 12, 13), (8, 15, 17), (7, 24, 25), and (9, 40, 41)? Prove that your answer is correct.

11. Prove the portion of Theorem 3.2.1 (page 48) that pertains to complete residue systems.

***12.** Consider the following proposed "**Theorem 3.2.1′**": Suppose that $\{z_1, z_2, \ldots, z_n\}$ is a complete or reduced residue system modulo m. Assume further that the integer b is relatively prime to the modulus m. Then the set $\{b + z_1, b + z_2, \ldots, b + z_n\}$ is again a complete or, respectively, reduced residue system modulo m.

***(a)** Is this a theorem?

***(b)** Is that half of the above statement that relates to complete residue systems modulo m correct?

***(c)** Is that half of the above statement that relates to reduced residue systems modulo m correct?

Prove that your answers to parts (a), (b), and (c) are correct.

***13.** True or false?

***(a)** $2^6 \equiv 1 \pmod 7$.

***(b)** $|(2^3 - 1)/7|_p \leqslant 1$ for all primes p.

***(c)** $3^n \equiv 1 \pmod{10}$ requires $n \geqslant 4$.

***(d)** $|(b^{p^{e-1}(p-1)} - 1)/p^e|_p \leqslant 1$ for all primes p and for all positive integers e.

***(e)** $4^{18} = 1 \pmod{27}$.

***(f)** $\phi(p^e) = p^e - p^{e-1}$ whenever p is a prime and e is a positive integer.

Justify your answers to parts (a)–(f).

14. We define the **Pell sequence** $\{P_n\}_{n=0}^{+\infty}$ as follows: $P_0 = 0$, $P_1 = 1$, and $P_{n+2} = 2P_{n+1} + P_n$ for all integers $n \geqslant 0$. Let $x_n = (P_{n+1})^2 - (P_n)^2$, $y_n = 2P_{n+1}P_n$, and $z_n = (P_{n+1})^2 + (P_n)^2$. Prove that for every positive integer n the numbers x_n, y_n, and z_n form a Pythagorean triple and, further, x_n and y_n are consecutive integers.

3.3 Linear and Quadratic Congruences

The notion of linear congruence first cropped up—and in a very natural manner—in the proof of Wilson's Theorem (see page 41). Problems 1, 8, and 9 (on pages 42 and 44) and Corollary 3.2.2 (on page 50) are other places where we have encountered linear congruences. The plan in this section is to study linear congruences somewhat more systematically and to introduce congruences of the second degree.

A **linear** (in the symbol X) **congruence** is anything of the form

$$bX \equiv c \pmod{m} \tag{1}$$

where b, c, and $m > 1$ are integers. Here are some natural questions to ask concerning linear congruence (1):

(a) Under what circumstances does there exist an integer x for which $bx \equiv c \pmod{m}$—i.e., when is our linear congruence solvable?

(b) If linear congruence (1) is solvable, then how many "essentially distinct" solutions does the congruence have? (The phrase "essentially distinct" will have to be made precise in order for question (b) to make sense.)

(c) How could we systematically go about finding all possible "essentially distinct" solutions to a given linear congruence?

We shall give complete answers to questions (a) and (b), and we will build up our intuition about question (c) via numerical examples and problems. An authoritative reference for a complete solution to question (c) is Niven and Zuckerman's *Introduction to the Theory of Numbers*.

Let's start by going after question (a). The congruence statement $bx \equiv c \pmod{m}$ is equivalent to the divisibility statement $m|(bx - c)$, which in turn is equivalent to the equation $bx - c = (-y)m$—or in its final form, $bx + my = c$. So the original linear congruence $bX \equiv c \pmod{m}$ is solvable if and only if there exist integers x and y for which $bx + my = c$. However, this is (by now) an ancient friend! The solvability of linear congruence (1) is equivalent to the requirement that the integer c be an <u>integral linear combination</u> of the input integers b and m. This requirement is, in turn, equivalent to the requirement that c be an integer multiple of the greatest common divisor of b and m, (b, m). So far, we have learned that the linear congruence $bX \equiv c \pmod{m}$ is solvable if and only if $(b, m)|c$. This is an extremely satisfactory state of affairs: we have a straightforward, efficient way of deciding whether or not a given linear congruence has any solutions. On to question (b)!

Suppose x_0 is a particular solution of our linear congruence, so that we have $bx_0 \equiv c \pmod{m}$. Then every integer of the form $x = x_0 + km$, regardless of the choice of the integer k, is also a solution of our linear congruence (why?). However, these aren't essentially distinct modulo m from the "base" solution x_0 because $x = x_0 + km \equiv x_0 \pmod{m}$, regardless of the choice of the integer k.

This fact motivates us to say that two solutions x_1 and x_2 of the linear congruence $bX \equiv c \pmod{m}$ are to be counted as different if and only if $x_1 \not\equiv x_2 \pmod{m}$. We formalize this in the following definition.

Definition 3.3.1: The **number of solutions** of the linear congruence $bX \equiv c$ (mod m) is the number of solutions of the given congruence that are pairwise incongruent modulo m.

Now that the meaning of "essentially distinct" solutions of a linear congruence has been spelled out, question (b) makes sense, and we can return to it.

We start by assuming that $G = (b, m)|c$ (why?). The congruence $bX \equiv c$(mod m) is equivalent (see page 40, part 4 of Theorem 3.1.2, and page 43, problem 7(c)) to the congruence $(b/(b, m))X \equiv c/(b, m)$(mod $m/(b, m)$), which, for convenience, may be rewritten as $(b/G)X \equiv c/G \pmod{m/G}$. I claim that the congruence $(b/G)X \equiv c/G \pmod{m/G}$ has exactly one solution, according to Definition 3.3.1, and that our original congruence has $G = (b, m)$ solutions—thus answering question (b). Here are the details: our assumption $G = (b, m)|c$ guaranteed the solvability of $(b/G)X \equiv c/G \pmod{m/G}$. If x_1 and x_2 both satisfy this congruence, then $(b/G)x_1 \equiv c/G \pmod{m/G}$ and $(b/G)x_2 \equiv c/G \pmod{m/G}$, so that $m/G|\{((b/G)x_1 - c/G) - ((b/G)x_2 - c/G)\} = (b/G) \cdot (x_1 - x_2)$, and so $x_1 \equiv x_2 \pmod{m/G}$ (why?). The point is that (according to Definition 3.3.1) x_1 and x_2 don't count as distinct solutions to this congruence because they are congruent to each other modulo m/G. The argument thus far has established the existence and uniqueness of the solution—call it x_0—of the congruence $(b/G)X \equiv (c/G) \pmod{m/G}$. Now consider the set $\{x_0, x_0 + m/G, x_0 + (2 \cdot m/G), \ldots, x_0 + ((G - 1) \cdot m/G)\}$. Every element of this set satisfies the congruence $(b/G)X \equiv (c/G) \pmod{m/G}$, in spite of only counting as one solution of that congruence. Hence, every element of this set satisfies our original linear congruence $bX \equiv c \pmod{m}$ because of the equivalence of these two congruences. However, $x_0 + (i \cdot m/G) \not\equiv x_0 + (j \cdot m/G) \pmod{m}$ if $0 \leqslant i < j \leqslant (G - 1)$, and so the congruence $bX \equiv c \pmod{m}$ is seen to have at least $G = (b, m)$ solutions. It remains to be shown that this is the full count—i.e., that the congruence $bX \equiv c \pmod{m}$ has precisely $G = (b, m)$ solutions. How would you argue for that? (See page 58, problem 10.)

With regard to question (c)—that of devising a reasonably efficient algorithm for obtaining the entire solution set of a given linear congruence: Niven and Zuckerman's *Introduction to the Theory of Numbers* has a readable discussion, complete with illustrative numerical examples, starting on page 41.

Speaking of illustrative numerical examples, let's show that we have already—by answering of questions (a) and (b)—significantly enhanced our ability to deal with linear congruences.

Example 3.3.1: The linear congruence $9X \equiv 13 \pmod{15}$ (see page 42, problem 1(f)) can now be seen immediately to have no solutions because $(9, 15) \nmid 13$. The linear congruence $9X \equiv 12 \pmod{15}$ (see page 42, problem 1(e)) is solvable because $3 = (9, 15) | 12$, and it has precisely three solutions. The congruence $(9/3)X \equiv (12/3) \pmod{15/3}$ has the solution $x_0 = 3$, from which we easily find that $\{3, 8, 13\}$ constitutes a full set of solutions for the original congruence.

Example 3.2.2: In the proof of Wilson's Theorem (page 41), we encountered the problem of establishing the existence and uniqueness of the integer j satisfying $ij \equiv 1 \pmod{p}$, together with the additional constraint $i < j \leqslant (p - 2)$. Here the integer i was any arbitrary but fixed integer satisfying $2 \leqslant i \leqslant (p - 2)$. With question (b) settled, it is obvious that each linear congruence of the form $iX \equiv 1 \pmod{p}$, $2 \leqslant i \leqslant (p - 2)$, has exactly one solution—call it j—and that we may assume that $0 < j < p$. Thus we could now shorten the proof of Wilson's Theorem considerably.

The situation for quadratic (i.e., second-degree in X) congruences is already substantially more difficult than is that for linear congruences. In Chapter 4 we will outline an approach to the problem of solving polynomial congruences of arbitrary degree, but we won't obtain a fully satisfactory answer to the problem of quadratic congruences until the final chapter of the text (Chapter 5). In that final chapter we shall establish Carl Friedrich Gauss' famous Law of Quadratic Reciprocity—the last major result to be discussed in this text and, I hope, a fitting climax for a brief and introductory account of the subject. In the meantime we shall content ourselves with learning some of the more accessible facts concerning quadratic congruences.

Consider a congruence of the form

$$bX^2 + cX + d \equiv 0 \pmod{m} \tag{2}$$

where, without loss of generality (why?), $b \not\equiv 0 \pmod{m}$. The easy case turns out to be that in which $(2b, m) = 1$, and we shall now make that assumption. Then our congruence is equivalent[1] to the congruence $4b(bX^2 + cX + d) \equiv 4b(0) \pmod{m}$, which may be rewritten in the form $(2bX + c)^2 \equiv c^2 - 4bd \pmod{m}$. Algebraically, we have completed the square on X. I now claim that it is enough, in our admittedly special case $(2b, m) = 1$, to be able to handle quadratic congruences of the (seemingly) very specialized form $Y^2 \equiv B \pmod{m}$. Here, Y corresponds to the quantity $2bX + c$, and B corresponds to the quantity $c^2 - 4bd$, which happens to be the discriminant of the original quadratic polynomial $bX^2 + cX + d$. Let's look at a concrete numerical example in order to get some feeling for what is going on.

[1] i.e., "has the same solution set as."

Example 3.3.3: Consider the quadratic congruence $2X^2 + 7X - 10 \equiv 0 \pmod{11}$, so that $b = 2$, $c = 7$, $d = -10$, $m = 11$, and the condition $(2b, m) = 1 = (2 \cdot 2, 11) = (4, 11) = 1$ is satisfied. The equivalent congruence $(2bX + c)^2 \equiv c^2 - 4bd \pmod{m}$ takes the form $(4X + 7)^2 \equiv 7^2 - (4 \cdot 2 \cdot (-10)) \equiv 49 + 80 \equiv 129 \equiv 8 \pmod{11}$. We learn that the original congruence has no solutions by checking that $Y^2 \equiv 8 \pmod{11}$ has no solutions. The point here is that the original problem can be replaced by an equivalent but easier-to-solve problem.

PROBLEMS

Solutions (or hints at solutions) to the problems marked with an asterisk (*) are given on page 126.

*1. In each of the following, use the criterion developed in this section to discover whether the given linear congruence is solvable:
   (a) $10X \equiv 30 \pmod{15}$
   ***(b)** $15X \equiv 10 \pmod{30}$
   (c) $30X \equiv 15 \pmod{10}$
   ***(d)** $64X \equiv 83 \pmod{105}$
   (e) $18X \equiv 21 \pmod{15}$
   (f) $18X \equiv 21 \pmod{24}$

*2. Characterize (precisely) the integers that satisfy the given congruence:
   (a) $10X \equiv 30 \pmod{15}$
   (b) $21X \equiv 18 \pmod{24}$
   ***(c)** $64X \equiv 83 \pmod{105}$
   ***(d)** $24X \equiv 18 \pmod{21}$

   [HINT: If there are any integers that satisfy the congruence in question, then they will constitute an arithmetic progression—see page 43, problem 6.]

*3. Determine the number of solutions (according to Definition 3.3.1, page 55) of each of the following:
   ***(a)** $24X \equiv 18 \pmod{21}$
   (b) $83X \equiv 64 \pmod{105}$
   (c) $10X \equiv 30 \pmod{15}$
   (d) $101X \equiv 229 \pmod{89}$
   ***(e)** $144X \equiv 36 \pmod{210}$

*4. Make a complete list of pairwise incongruent solutions for each of the five linear congruences of problem 3.

5. If the integer b is selected at random from the set $\{1, 2, 3, \ldots, 12\}$ and the integer c is selected at random from the set $\{1, 2, 3, \ldots, 17\}$, what

is the probability that the resulting linear congruence $bX \equiv c \pmod{12}$ has exactly one solution? At most one solution?

***6.** In each of the following quadratic congruences, list all the values of x satisfying both the given congruence and the inequalities $0 < x < (m - 1)$, with m the modulus of the congruence.

***(a)** $X^2 \equiv 4 \pmod 5$
(b) $X^2 \equiv 1 \pmod 8$
(c) $2X^2 \equiv 3 \pmod{12}$
***(d)** $X^2 \equiv -1 \pmod 7$
(e) $2X^2 \equiv 4 \pmod{12}$
(f) $X^2 \equiv -1 \pmod{11}$

***7.** If we agree to count the number of solutions of a quadratic congruence according to the definition made for linear congruences (Definition 3.3.1), then determine the number of solutions of each of the congruences of problem 6.

***8.** In each of the following, use the completing-the-square technique illustrated in Example 3.3.3 on page 57 to produce a congruence of the form $Y^2 \equiv B \pmod m$ that is equivalent to the given congruence; then determine whether the given congruence is solvable.

(a) $X^2 + X + 7 \equiv 0 \pmod{27}$
(b) $3X^2 + 6X + 5 \equiv 0 \pmod 7$
(c) $X^2 + X + 47 \equiv 0 \pmod{343}$
***(d)** $4X^2 + X \equiv 14 \pmod{13}$

***9.** Assume that p is a prime. Prove that $X^2 \equiv -1 \pmod p$ is solvable if and only if $p = 2$ or $p \equiv 1 \pmod 4$. [HINTS: (1) If p is an odd prime, pair off the factors of $(p - 1)!$ by letting r correspond to $p - r$ for $1 \leqslant r \leqslant (p - 1)/2$. (2) If $p \equiv 1 \pmod 4$, use Wilson's Theorem (see page 41) to complete the proof. (3) If $p \not\equiv 1 \pmod 4$ make a proof by contradiction, with the help of Fermat's Little Theorem (see page 50).]

10. On the assumption that $G = (b, m)|c$, prove that the linear congruence $bX \equiv c \pmod m$ has no more than G solutions, thus completing the argument given on page 55.

3.4 The Chinese Remainder Theorem

In Section 3.3 we learned that the (single) linear congruence $bX \equiv c \pmod m$ is solvable if and only if $G = (b, m)|c$. Furthermore, if this condition is satisfied, the original congruence is equivalent to the congruence $(b/(b, m))X \equiv (c/(b, m)) \pmod{m/(b, m)}$. This new congruence could be rewritten as $BX \equiv C$ (mod

M), upon defining $B = b/(b, m)$, $C = c/(b, m)$, and $M = m/(b, m)$. Notice that $(B, M) = (b/(b, m), m/(b, m)) = 1$, which means that there must exist integers u and v for which $Bu + Mv = 1$. I claim that the congruence $BX \equiv C$ (mod M) is equivalent to the congruence $(Bu)X \equiv (Cu)$ (mod M) (why?). However, $Bu \equiv 1$ (mod M), and so finally our original congruence $bX \equiv c$ (mod m) has been replaced by a congruence of the form $X \equiv D$ (mod M) for suitable choices of D and M. The point of all this is that, in studying linear congruences, we need only consider those of the form $X \equiv b$ (mod m). As an illustration, consider the following example.

Example 3.4.1: The linear congruence $144X \equiv 36$ (mod 210) is equivalent to the simpler congruence $X \equiv 9$ (mod 35). Please check the details for yourself (see also page 57, problem 3(e)).

Now that we have a pretty good understanding of the situation for a single linear congruence, we tackle the problem of determining when a system of finitely many simultaneous linear congruences has solutions. The Chinese Remainder Theorem gives necessary and sufficient conditions for such a system to be solvable. We now state and prove this important result.

Theorem 3.4.1 – The Chinese Remainder Theorem: The system

$$\begin{aligned} X &\equiv b_1 \pmod{m_1} \\ X &\equiv b_2 \pmod{m_2} \\ &\ \ \vdots \\ X &\equiv b_n \pmod{m_n} \end{aligned}$$

of **simultaneous linear congruences** is solvable if and only if $(m_i, m_j) | (b_i - b_j)$ for every pair of subscripts i and j satisfying $1 \leqslant i < j \leqslant n$. Any two solutions of the system are congruent modulo the least common multiple of the moduli $[m_1, m_2, \ldots, m_n]$.

Proof: The necessity part of the argument is the easier part, and we deal with it first. In order that the system as a whole be solvable, it is certainly necessary that every pair—generically, $X \equiv b_i$ (mod m_i) and $X \equiv b_j$ (mod m_j), where $1 \leqslant i < j \leqslant n$—be solvable. Now $x \equiv b_i$ (mod m_i) if and only if $x = b_i + zm_i$ for some integer z. The second requirement, that $x \equiv b_j$ (mod m_j), now takes the form $b_i + zm_i \equiv b_j$ (mod m_j) or $m_i z \equiv b_j - b_i$ (mod m_j). However, we know (see page 54) that such a linear (in z) congruence is solvable if and only if the greatest common divisor $G = (m_i, m_j)$ is a divisor of $b_j - b_i$, and so the necessity of the conditions $(m_i, m_j)|(b_i - b_j)$ is established.

We now assume that $(m_i, m_j)|(b_i - b_j)$ for every pair of subscripts i and j satisfying $1 \leq i < j \leq n$—i.e., we assume that every pair of our congruences is solvable—and we argue that the entire system therefore has a solution. We must show the following (why?): if any pair of the original list of n congruences is solved to produce a single new congruence, then the new list of $(n - 1)$ congruences consisting of this new congruence together with the remaining $(n - 2)$ congruences from the original list still has the property that every pair of congruences (from the new list) is solvable. We already know that the pair of congruences, $X \equiv b_1 \pmod{m_1}$ and $X \equiv b_2 \pmod{m_2}$, is equivalent to the single congruence $m_1 z \equiv b_2 - b_1 \pmod{m_2}$. This congruence has (m_1, m_2) solutions modulo m_2, but z is uniquely determined modulo $m_2/(m_1, m_2)$, and so $x = b_1 + zm_1$ (recalling our notational setup at the beginning of the proof) is uniquely determined modulo $m_1 m_2/(m_1, m_2) = [m_1, m_2]$. Let us write $x \equiv b_{12} \pmod{[m_1, m_2]}$, where b_{12} is uniquely determined modulo $[m_1, m_2]$.

The rest of our job is to prove that every pair of congruences of the form $X \equiv b_{12} \pmod{[m_1, m_2]}$ and $X \equiv b_i \pmod{m_i}$, where $3 \leq i \leq n$ is solvable. This requires precisely that $(m_i, [m_1, m_2])|(b_i - b_{12})$. In valuation-theoretic terms, this requirement becomes $|(b_i - b_{12})/(m_i, [m_1, m_2])|_p \leq 1$ for all i satisfying $3 \leq i \leq n$ and for all primes p. Let us define α_j, for $1 \leq j \leq n$, by $p_i^{\alpha_j} \| m_j$. Then $p^{\max\{\alpha_1, \alpha_2\}} \| [m_1, m_2]$ (refer to page 24, if necessary) and $p^{\min\{\alpha_i, \max\{\alpha_1, \alpha_2\}\}} \| (m_i, [m_1, m_2])$; so it suffices to prove that this same power of p divides the numerator $(b_i - b_{12})$. Now $b_i - b_{12} = (b_1 - b_{12}) - (b_1 - b_i)$ and $p^{\alpha_1}|(b_1 - b_{12})$, while $p^{\min\{\alpha_1, \alpha_i\}}|(b_1 - b_i)$ (why?); so $p^{\min\{\alpha_1, \alpha_i\}}|(b_i - b_{12})$. Likewise, $p_i^{\min\{\alpha_2, \alpha_i\}}|(b_i - b_{12})$. Hence, $p_i^{\max\{\min\{\alpha_1, \alpha_i\}, \min\{\alpha_2, \alpha_i\}\}}|(b_i - b_{12})$. Since $\max\{\min\{r, t\}, \min\{s, t\}\} = \min\{t, \max\{r, s\}\}$ (see problem 10 on page 23) is true regardless of the choice of the real numbers r, s, and t, $(m_i, [m_1, m_2])|(b_i - b_{12})$ has been established for all subscripts i satisfying $3 \leq i \leq n$, and the sufficiency of the conditions $(m_i, m_j)|(b_i - b_j)$, $1 \leq i < j \leq n$, has been demonstrated.

Concerning the uniqueness of the solution of the system mod $[m_1, m_2, \ldots, m_n]$, we already know that b_{12} is uniquely determined mod $[m_1, m_2]$. Solving $X \equiv b_{12} \pmod{[m_1, m_2]}$ and $X \equiv b_3 \pmod{m_3}$ leads to $x \equiv b_{123} \pmod{[m_1, m_2, m_3]}$, where b_{123} is uniquely determined modulo $[m_1, m_2, m_3]$. Iterating this procedure gives the result.

Example 3.4.2: Professor Avery got a sabbatical in 1975 and now gets another sabbatical every seven years. He is also in the habit of buying a new car every five years, having first done this in 1976. Could it happen that he buys a new car while on sabbatical during a leap year?

A special, but very important, case of the Chinese Remainder Theorem is the following corollary.

Corollary 3.4.1: The system

$$X \equiv b_1 \pmod{m_1}$$
$$X \equiv b_2 \pmod{m_2}$$
$$\vdots$$
$$X \equiv b_n \pmod{m_n}$$

of simultaneous linear congruences is solvable if the moduli are pairwise relatively prime. Any two solutions of the system are congruent modulo the product of the moduli.

Proof: Obvious.

Whereas the Chinese Remainder Theorem itself is an existence (and uniqueness, of a sort) result, we can make the proof constructive in the case of the corollary; i.e., we can actually give a formula that will provide a particular solution of the system. We start by noticing that each congruence of the form $m/m_iX \equiv 1 \pmod{m_i}$, $1 \leqslant i \leqslant n$, is solvable because $(m/m_i, m_i) = 1$. (Here, $m = m_1 \cdot m_2 \cdot \ldots \cdot m_n$ because the moduli are assumed to be pairwise relatively prime.) Let us write $m/m_ic_i \equiv 1 \pmod{m_i}$ for each available value of i; i.e., we are letting c_i denote a solution of the congruence $m/m_iX \equiv 1 \pmod{m_i}$. Notice that $m/m_ic_i \equiv 0 \pmod{m_j}$ for all $j \neq i$ because $j \neq i$ implies $m_j | m/m_i$. Now consider $x_0 = \sum_{i=1}^{n} (m/m_i)c_ib_i$. Regardless of the choice of the modulus m_j, $1 \leqslant j \leqslant n$, we get $x_0 \equiv \sum_{i=1}^{n} (m/m_i)c_ib_i \equiv (m/m_j)c_jb_j \equiv b_j \pmod{m_j}$, so that x_0 is a solution of the original system.

Example 3.4.3: Let's find the smallest positive integer that leaves a remainder of, respectively, 1, 2, and 3, upon division by 3, 4, and 5. We start by finding *a* solution of the system:

$$X \equiv 1 \pmod{3}$$
$$X \equiv 2 \pmod{4}$$
$$X \equiv 3 \pmod{5}.$$

For example, we may take $c_1 = -1$, $c_2 = 3$, and $c_3 = 3$ (please check), and so *a* solution x_0 is $20(-1)(1) + 15(3)(2) + 12(3)(3) = 178$. Hence the answer is 58, because any two solutions of this particular system differ by an integral multiple of 60.

Remarks:

1. The terminology isn't uniform; a lot of people refer to Corollary 3.4.1 (page 61) as the Chinese Remainder Theorem.
2. The set of all integers x satisfying the single linear congruence $X \equiv b \pmod{m}$ is precisely the set $\{\ldots, b - 2m, b - m, b, b + m, b + 2m, b + 3m, \ldots\}$, and thus is the arithmetic progression containing the integer b and having common difference m (see page 43, problem 6). The Chinese Remainder Theorem can therefore be reinterpreted as follows: the intersection of a finite number of arithmetic progressions is nonempty if and only if each pair of these arithmetic progressions overlaps.
3. The Chinese Remainder Theorem can be generalized out of sight and is an extremely important theoretical tool in advanced work.

PROBLEMS

Solutions (or hints at solutions) to the problems marked with an asterisk (*) are given on page 127.

*1. In each of the following, determine whether the given pair of linear congruences can be solved simultaneously. Find all integers x (if any) satisfying $0 < x \leqslant 100$ and also satisfying the given pair of congruences.
   (a) $X \equiv 2 \pmod{6}$
   $X \equiv 3 \pmod{8}$
   (b) $X \equiv 2 \pmod{6}$
   $X \equiv 3 \pmod{7}$
   *(c) $5X \equiv -2 \pmod{10}$
   $X \equiv 1 \pmod{4}$
   *(d) $5X \equiv 6 \pmod{7}$
   $4X \equiv 5 \pmod{9}$

*2. Prove that $X \equiv b \pmod{m}$ is solvable if and only if $X \equiv b \pmod{p^e}$ is solvable for every p-component (see page 23, problem 8) p^e of m. (Of course, $X \equiv b \pmod{m}$ is solvable!)

*3. Find a solution of the congruence $19X \equiv 1 \pmod{140}$, using Corollary 3.4.1 in conjunction with the idea underlying the previous problem.

*4. Monkey-Sailor-Coconut Problem. Three sailors pick n coconuts and retire for the night. One of the sailors gets up in the middle of the night and, not trusting the other two sailors, attempts to take his "fair share," namely $n/3$ coconuts. However, upon counting the coconuts he finds that n isn't divisible by 3; rather, n is 1 more than an integral multiple of 3. He decides to give the extra coconut to the monkey and then takes (as he now can) one-third of the remaining pile of coconuts. The second sailor gets up after the first sailor has operated on the original pile of n coconuts as described, and he, too, tries to take his fair share; but he discovers, as did the first sailor, that the (now somewhat smaller) pile of coconuts he faces contains a number of coconuts that is 1 more than a multiple of 3. He gives the extra coconut to the monkey and then takes one-third of the remaining coconuts. The third sailor goes through exactly the same ritual, so that the monkey now has three coconuts. In the morning the three sailors approach the (by now) substantially diminished pile of coconuts and—upon trying to divide the pile fairly—find themselves presenting the monkey with his fourth and final coconut. What is the *smallest* value of n for which the foregoing situation could have occurred? [HINTS: (1) $n = -2$ is, in some sense, the optimal mathematical solution, but it has physical drawbacks. (2) Each of four distinct positive integers must be divisible by 3, the first of these four being $n - 1$.] Solve the analogous problem for five sailors.

5. Assume that b, c, and n are integers with $(b, c) = 1$. Prove that there exists an integer x for which $(bx + c, n) = 1$. [HINTS: (1) If p is any prime divisor of n, it is possible to choose the integer x in such a way that $p \nmid (bx + c)$. (2) Apply the Chinese Remainder Theorem.]

***6.** Determine precisely which leap years, if any, are "good" for Professor Avery, in the sense that he will both have a new car and be on sabbatical. (See Example 3.4.2, page 60.)

7. Prove that $x \equiv b \pmod{m}$ if and only if $|(x - b)/p^e|_p \leqslant 1$ for all primes p dividing m, where, furthermore, it is to be understood that $p^e \| m$; i.e., $p^e | m$, but $p^{e+1} \nmid m$. (Compare page 62, problem 2.)

8. An integer $n \neq 0, \pm 1$ is said to be **squarefree** if its canonical decomposition is of the form $n = \pm p_1 \cdot p_2 \cdot \ldots \cdot p_k$; i.e., no prime appears to a power higher than 1. Prove that there exist arbitrarily long blocks of consecutive integers, no one of which is squarefree.

9. Prove that there exist arbitrarily long blocks of consecutive integers, no one of which is prime.

10. Prove that $\max\{\min\{r, t\}, \min\{s, t\}\} = \min\{t, \max\{r, s\}\}$, regardless of the choice of the real numbers r, s, and t.

***11.*(a)** Does the result of problem 8 imply the result of problem 9?
***(b)** Does the result of problem 9 imply the result of problem 8?

3.5 More on the Euler φ-Function, and the Order Symbol

Having studied the Chinese Remainder Theorem in the last section, we immediately illustrate its power as a theoretical tool by applying it to help us learn more about the Euler φ-function (see Definition 3.2.3 on page 48). We begin by proving the following theorem.

Theorem 3.5.1: $\phi(mn) = \phi(m)\phi(n)$ for all choices of relatively prime positive integers m and n.

Proof: Let $R = \{r_1, r_2, \ldots, r_{\phi(m)}\}$, $S = \{s_1, s_2, \ldots, s_{\phi(n)}\}$, and $T = \{t_1, t_2, \ldots, t_{\phi(mn)}\}$ denote, respectively, reduced residue systems modulo m, n, and mn. Since the Cartesian product set $R \times S = \{(r_i, s_j) | 1 \leq i \leq \phi(m), 1 \leq j \leq \phi(n)\}$ contains $\phi(m)\phi(n)$ elements, we can complete the proof by showing that T contains the same number of elements as does $R \times S$. If t_k is any element of T, then $(t_k, mn) = 1$, which tells us that $(t_k, m) = (t_k, n) = 1$. Since $(t_k, m) = 1$, we must have $t_k \equiv r_i \pmod{m}$ for some subscript i satisfying $1 \leq i \leq \phi(m)$, and furthermore there can be only one such subscript (see page 51, problem 7(a)). Likewise, $t_k \equiv s_j \pmod{n}$ for some uniquely determined integer j satisfying $1 \leq j \leq \phi(n)$. Hence, the rule that sends t_k to the ordered pair (r_i, s_j) is a function from T into $R \times S$. Moreover, distinct elements of T get sent to distinct elements of $R \times S$ under this mapping, and so $^{\#}(T) \leq {}^{\#}(R \times S)$, where the symbol $^{\#}(A)$ denotes the number of elements of the set A. (Please verify the two statements made in the last sentence; see problem 24 on page 70.) If we can play the same game once more, so as to invent an appropriate function from $R \times S$ back into T, we will be able to conclude that $^{\#}(R \times S) \leq {}^{\#}(T)$, and so be done (why?). It is precisely at this point that the Chinese Remainder Theorem comes to our aid. If we preassign the ordered pair (r_i, s_j)—a generic element of $R \times S$—then there is one and, modulo mn, only one integer t_k that satisfies the two congruences

$$X \equiv r_i \pmod{m}$$
$$X \equiv s_j \pmod{n}.$$

Since $t_k \equiv r_i \pmod{m}$ and since $(r_i, m) = 1$, it follows that $(t_k, m) = 1$. Likewise, since $t_k \equiv s_j \pmod{n}$ and $(s_j, n) = 1$, we get $(t_k, n) = 1$. Since $(t_k, m) = 1$ and $(t_k, n) = 1$, we know that $(t_k, mn) = 1$ (by property 5 given on page 10), so that t_k is an element of T; and thus we have our "backward" (technically, *inverse*) function from $R \times S$ into T.

Remark: A function f whose natural domain of definition is the set of positive integers is said to be **multiplicative** (some authors use the term **weakly multiplicative**) if $f(mn) = f(m)f(n)$ for all choices of relatively prime positive integers m and n. If f is such a function and if we can discover a formula for $f(p^e)$ where p^e takes on all pure prime power values, then we can obtain a formula for $f(n)$ for every positive integer n (why?). (See also note 6 on page 78.)

We apply this general truth to the specific case of the Euler ϕ-function to obtain the following result.

Theorem 3.5.2: If the positive integer n has the canonical decomposition $p_1^{e_1} \cdot p_2^{e_2} \cdot \ldots \cdot p_k^{e_k} = \prod_p p^e$, then

$$\begin{aligned}\phi(n) &= p_1^{e_1-1} \cdot p_2^{e_2-1} \cdot \ldots \cdot p_k^{e_k-1} \cdot (p_1 - 1) \cdot (p_2 - 1) \cdot \ldots \cdot (p_k - 1)\\ &= \prod_p p^{e-1}(p-1)\end{aligned}$$

where, in each case, the product is understood to extend over all the distinct prime divisors of n.

Proof: In light of the remark immediately preceding the statement of our theorem, it suffices that we come up with a formula for $\phi(p^e)$, where p^e is a completely arbitrary pure prime power—i.e., where p is an arbitrary prime and e is an arbitrary positive integer. According to Definition 3.2.3 on page 48, $\phi(p^e)$ is the number of positive integers in the interval $[1, p^e]$ that are relatively prime to p^e. As in probability, it is often easier for us to attack the complementary event—in this case, counting the number of positive integers in the interval $[1, p^e]$ that fail to be relatively prime to p^e. An integer z fails to be relatively prime to p^e if and only if $(z, p^e) > 1$ or, equivalently, if and only if $(z, p) > 1$, or, finally, if and only if $p|z$. Hence, the integers in the interval $[1, p^e]$ that fail to be relatively prime to p^e are precisely all integers in the interval that are multiples of p—namely, $1 \cdot p, 2 \cdot p, \ldots, p^{e-1} \cdot p$. The number of integers in this list is p^{e-1}, and hence $\phi(p^e) = p^e - p^{e-1} = p^{e-1}(p - 1)$. Now $\phi(n) = \phi\left(\prod_p p^e\right) = \prod_p \phi(p^e) = \prod_p p^{e-1}(p - 1)$, using the multiplicativity of ϕ.

Example 3.5.1: $\phi(24) = \phi(2^3 \cdot 3^1) = \phi(2^3) \cdot \phi(3^1) = 4 \cdot 2 = 8$. Note, however, that $\phi(4) \cdot \phi(6) = 2 \cdot 2 = 4 \neq 8$ even though $4 \cdot 6 = 24$.

We introduce the concept that pervades the latter portion of this section by means of a couple of easy, yet instructive numerical examples.

Example 3.5.2: If $b = 11$ and $m = 3$, does there exist a positive integer n for which $b^n \equiv 1 \pmod{m}$—i.e., $11^n \equiv 1 \pmod 3$—and, if so, what is the smallest value that we can choose for n? A straightforward, albeit naive, approach to this problem is to try, successively, $n = 1, 2, 3, 4, \ldots$. Now, $11^1 \not\equiv 1 \pmod 3$ because $3 \nmid (11 - 1) = 10$, and so $n \neq 1$. However, $3|(11^2 - 1) = 120$, so $11^n \equiv 1 \pmod 3$ has solutions in positive integers, and $n = 2$ is the least positive integral solution.

Example 3.5.3: Does the exponential congruence $5^n \equiv 1 \pmod 7$ have any positive integral solutions? If so, what is the least among them? Since $5^1 \not\equiv 1 \pmod 7$, $5^2 = 25 \equiv 4 \not\equiv 1 \pmod 7$, $5^3 = 5^1 \cdot 5^2 \equiv 5 \cdot 4 = 20 \equiv -1 \not\equiv 1$

(mod 7), $5^4 = (5^2)^2 \equiv 4^2 \equiv 2 \not\equiv 1 \pmod 7$, $5^5 = 5^2 \cdot 5^3 \equiv 4(-1) = -4 \not\equiv 1 \pmod 7$, and $5^6 = (5^3)^2 \equiv (-1)^2 = 1 \equiv 1 \pmod 7$, we see—but only by brute force—that $n = 6$ is the least positive integral solution of $5^n \equiv 1 \pmod 7$.

Now let's try to analyze the general situation of the exponential congruence $b^n \equiv 1 \pmod m$, where it is understood that b and m are preassigned integers and n is the unknown integral variable. We know that we may assume $m > 1$ without any loss of generality, so we start by making that assumption. Next, while $b^0 \equiv 1 \pmod m$ is invariably true, it is equally invariably uninteresting; so let us always assume that n is nonzero. In fact, let us assume for the time being that n is positive, so that—if there are any solutions—there is a least solution. One last observation is that $(b, m) = 1$ is necessary if $b^n \equiv 1 \pmod m$ is to have any solutions (why?). Having peeled off the layers obscuring the core of this problem, we now realize that the Euler-Fermat Theorem (see page 49) guarantees that $b^{\phi(m)} \equiv 1 \pmod m$, and so the existence of positive integral solutions of $b^n \equiv 1 \pmod m$ is established. It is time to make the most important definition of this section.

Definition 3.5.1: The least positive integral solution of the exponential congruence $b^n \equiv 1 \pmod m$ for given integers b and $m > 1$ satisfying $(b, m) = 1$ is denoted by $\mathbf{ord}_m\boldsymbol{b}$, which is read **the order of *b* modulo *m*.**

Remark: The numerical results of Examples 3.5.2 and 3.5.3 amount to saying that $\mathrm{ord}_3 11 = 2$ and $\mathrm{ord}_7 5 = 6$.

The minimality of the positive integer $\mathrm{ord}_m b$ makes it quite straightforward—using the laws of exponents and, once again, the Division Algorithm—to prove that $\mathrm{ord}_m b \mid \phi(m)$ (see page 68, problem 6). The special but extremely important case of equality (rather than simply divisibility) provokes our last definition of this section.

Definition 3.5.2: The integer b is said to be a **primitive root modulo *m*** if (and, of course, only if) $\mathrm{ord}_m b = \phi(m)$.

Remark: Thus 5 is a primitive root modulo 7, and 11 is a primitive root modulo 3, but 7 isn't a primitive root modulo 8.

An obvious question at this juncture is: "Which moduli possess primitive roots?" We have learned that $\mathrm{ord}_m b$ always exists (once the obvious necessary condition $(b, m) = 1$ is assumed), but there is no reason, a priori, to believe that each and every positive integer m has one or more primitive roots. I will ask you to prove that $m = 8$ has no primitive root, while $m = 7$ has exactly two primitive roots[1] (see problem 5 on page 68). Already we can notice a marked divergence between the subjects of our two definitions: given b and m with $(b, m) = 1$, $\text{ord}_m b$ always exists and is always unique; but, given m, there may not exist a primitive root modulo m, or if one does exist it may not be unique. Section 3.6 is given over to obtaining a complete characterization of moduli that possess primitive roots.

We now embark on a program I like to refer to as "the algebra of the order symbol"; this involves a systematic development of the basic properties of $\text{ord}_m b$. Let me start by making a list of some of the useful and accessible properties of $\text{ord}_m b$.

List of Some of the Basic Properties of $\text{ord}_m b$

1. $\text{ord}_m b | \phi(m)$
2. If $b^n \equiv 1 \pmod{m}$, then (and only then) $\text{ord}_m b | n$
3. $b^r \equiv b^s \pmod{m}$ if and only if $\text{ord}_m b | (r - s)$
4. $\text{ord}_m(b^k) = \text{ord}_m b/(k, \text{ord}_m b)$
5. $\text{ord}_m(bc) | \text{ord}_m b \cdot \text{ord}_m c$
6. $bc \equiv 1 \pmod{m}$ implies $\text{ord}_m b = \text{ord}_m c$

I will prove properties 4 and 5 and ask you to provide proofs for the other four properties listed.

Proof of Property 4: From property 2 of our list, we see that $(b^k)^n \equiv 1 \pmod{m}$ if and only if $\text{ord}_m b | kn$. However, this requirement that $\text{ord}_m b | kn$ is equivalent to the requirement that $\text{ord}_m b/(\text{ord}_m b, k) | \{k/(\text{ord}_m b, k)\} n$, which in turn is equivalent to the requirement that $\{\text{ord}_m b/(\text{ord}_m b, k)\} | n$, since $\text{ord}_m b/(\text{ord}_m b, k)$ is relatively prime to $k/(\text{ord}_m b, k)$. Hence the minimal available value of n is precisely $\text{ord}_m b/(\text{ord}_m b, k)$, and we are done.

Proof of Property 5: We have $(bc)^{\text{ord}_m b \cdot \text{ord}_m c} = (b^{\text{ord}_m b})^{\text{ord}_m c} \cdot (c^{\text{ord}_m c})^{\text{ord}_m b} \equiv 1^{\text{ord}_m c} \cdot 1^{\text{ord}_m b} \equiv 1 \pmod{m}$, and we are finished, upon invoking property 2 once more.

PROBLEMS

Solutions (or hints at solutions) to the problems marked with an asterisk (*) are given on page 127.

*1. (a) Determine the number of positive integers that are less than 10,000 and relatively prime to 10,000.

[1]Assume, without loss of generality, that $1 \leqslant b \leqslant m$—careful counting!

(b) Determine the number of positive integers that are less than or equal to 420 and relatively prime to $70 = 420/6$.

***(c)** If k and m are positive integers, find a formula representing the number of positive integers that are less than or equal to km and relatively prime to km.

***(d)** If k and m are positive integers, find a formula representing the number of positive integers that are less than or equal to km and relatively prime to m.

***(e)** Which of the two formulae you found in parts (c) and (d) represents the larger number?[1] Give a reason or proof justifying your answer.

(f) Characterize the ordered pairs (k, m) of positive integers for which the formulae you found in parts (c) and (d) give precisely the same answer.

***2.*(a)** Find all solutions x of $\phi(x) = 4$.

***(b)** Find all solutions x of $\phi(x) = 20$.

***3.** In each of the following, decide whether the statement is true or false. If it is true, provide a proof. If it is false, provide a "disproof."

***(a)** $\phi(mn) = \phi(n)\phi(m)$ for all choices of positive integers m and n.

***(b)** $\phi(n)$ is even for "almost all"[2] positive integers n.

***(c)** $\phi(n)\phi(m) > \phi(mn)$ for all choices of positive integers m and n satisfying the condition $(m, n) = 1$.

***(d)** $\phi(m)\phi(n) < \phi(mn)$ for all choices of positive integers m and n satisfying the condition $(m, n) > 1$.

***(e)** $\phi(n) = \phi(2n)$ whenever n is an odd positive integer.

***(f)** There are infinitely many positive integers n for which $3 \nmid \phi(n)$.

***(g)** $\phi(n) = 14$ has no solution.

***4.** Evaluate each meaningful symbol below.

(a) $\text{ord}_4 1$

(b) $\text{ord}_4 2$

***(c)** $\text{ord}_{1,093} 3$

(d) $\text{ord}_m 1$, for every $m > 1$.

5. Find all primitive roots (if any!) of $m = 7$. Find all primitive roots (if any!) of $m = 8$.

***6.** Prove that $\text{ord}_m b | \phi(m)$. [HINT: Use the Division Algorithm and the laws of exponents.]

***7.*(a)** Is 12 a primitive root of 7? Justify your answer.

***(b)** Is -5 a primitive root of 7? Justify your answer.

[1] Or at least no smaller a number, strictly speaking.

[2] i.e., for all but a finite number, in elementary number theory.

***(c)** If g is a primitive root of m and if $h \equiv g \pmod{m}$, then what can you conclude about h? Prove that your answer is correct.

***(d)** $b \equiv c \pmod{m}$ implies $\text{ord}_m b = \text{ord}_m c$? Proof?

***(e)** Is the converse of part (d) of this problem a valid implication? Justify your answer.

8. Establish properties 2, 3, and 6 of the order symbol, as listed on page 67.

***9.*(a)** If $d|n$, does it follow that $\phi(d)|\phi(n)$? Prove that your answer is correct.

***(b)** If $\phi(d)|\phi(n)$, does it follow that $d|n$? Prove that your answer is correct.

***10.** Prove that, for a fixed integer n, the equation $\phi(x) = n$ has only finitely many solutions x. [HINT: Consider $\lim_{x\to+\infty} \phi(x)$.]

***11.** Prove that, if $m|n$, then $\phi(mn) = m\phi(n)$.

12. Prove that, if every prime that divides m also divides n, then $\phi(nm) = m\phi(n)$.

***13.** Assume that n has k distinct odd prime factors.

***(a)** Does it follow that $2^k|\phi(n)$? Prove that your answer is correct.

***(b)** Does it follow that $2^k\|\phi(n)$? (Remember that $2^k\|\phi(n)$ means $2^k|\phi(n)$, but $2^{k+1}\nmid\phi(n)$.) Prove that your answer is correct.

***14.** Which of the following two problems is the more difficult to solve?

Problem 1: Characterize those positive integers n for which $\phi(n)|(n - 1)$.

Problem 2: Characterize those positive integers n for which $\phi(n)|n$.

"Prove" that your answer is correct by solving at least one of Problem 1 and Problem 2.

***15.** Assume that $\text{ord}_p b = 3$, with p being a prime. Conclude that $1 + b + b^2 \equiv 0 \pmod{p}$ and that $\text{ord}_p(1 + b) = 6$.

***16.** Show that, if p is an odd prime and $\text{ord}_{p^e}(b) = 2u$, then $b^u \equiv -1 \pmod{p^e}$. What about the case of $p = 2$?

17. Show that, if p is an odd prime and $b^u \equiv -1 \pmod{p}$, then $\text{ord}_p b = 2n$ for some positive integer n, and u is an odd multiple of n.

***18. Another primality test.** If there exists an integer b for which $b^{m-1} \equiv 1 \pmod{m}$, while $b^{(m-1)/p} \not\equiv 1 \pmod{m}$ regardless of the choice of the prime $p|(m - 1)$, then m is prime.

***19.** Assume that $\text{ord}_m b = h$ and that $\text{ord}_m c = k$.

(a) Prove that $\text{ord}_m bc|hk$.

***(b)** Prove that $\text{ord}_m(bc)|[h, k]$.

(c) Prove that $\text{ord}_m(bc) = hk$ if $(h, k) = 1$.

(d) Is it always correct to claim that $\text{ord}_m(bc) = hk$? Justify your answer.

***20.** Is it correct to claim that $\phi(mn) = n\phi(m)$ if and only if every prime divisor of n is also a prime divisor of m? Prove that your answer is correct.

21. Prove that, if P denotes the product of the primes common to m and n, then $\phi(mn) = P\phi(m)\phi(n)/\phi(P)$. In particular, $(m, n) > 1$ implies $\phi(mn) > \phi(m)\phi(n)$.

22. Prove that, if $(m, n) = G$, then $\phi(mn) = G\phi(n)\phi(m)/\phi(G)$. In particular, $(m, n) > 1$ implies $\phi(nm) > \phi(m)\phi(n)$. (Now look back at page 68, problem 3(d).)

23. Prove that the function $f(n) = n - \phi(n)$ is a strictly increasing arithmetic function, in the sense that—whenever $d|n$ and $d < n$—it follows that $f(d) < f(n)$.

24. Let f denote the function from T to $R \times S$, as defined on page 64 in the course of the proof of Theorem 3.5.1. Thus $f(t_k) = (r_i, s_j)$ for each element t_k of the set T.
    (a) Prove that f takes distinct elements of T to distinct elements of $R \times S$ (i.e, f is one-to-one, or injective).
    (b) Conclude, using the result in part (a) of this problem, that $^{\#}(T) \leqslant {}^{\#}(R \times S)$.

25. Prove that $\sum_{d|n} \phi(d) = n$, for each positive integer n.

26. If $g(n)$ denotes the sum of all positive integers that are less than or equal to n and relatively prime to n, prove that the function g is injective; i.e., prove that, if $g(n) = g(m)$, then $n = m$.

***27.** Define $f(n)$ to be the product of all the positive (integral) divisors of n. Does $f(m) = f(n)$ imply $m = n$? Prove that your answer is correct.

28. Prove that $n|\phi(b^n - 1)$ whenever $b > 1$.

29. Prove that $2n|\phi(b^n + 1)$ whenever $b > 1$.

***30.** Prove that $\text{ord}_{2^n}(3) = 2^{n-2}$ for all integers $n > 2$.

***31.** Does there exist an integer $m > 1$ for which $2^m \equiv -1$ or $-2 \pmod{m + 1}$? Proof?

***32.**†Prove that $\sum_{d|n} \phi(d)b^{n/d} \equiv 0 \pmod{n}$ for all positive integers b and n satisfying $(b, n) = 1$.

33. The positive integer 113 has the following properties: the sum of its digits is a prime; the product of its digits is a prime; and all the positive integers that can be obtained from it by permuting its decimal digits are prime—in this case 131 and 311 are both prime. Do any other positive integers satisfying all three of these properties exist? If so, find all of them, and prove that you have done so.

3.6 The Primitive Root Theorem

In Section 3.5, we defined what it means for an integer b to be a primitive root relative to a particular modulus $m > 1$. We found (page 68, problem 5) that some moduli—for instance $m = 8$—have no primitive roots, while other moduli—for instance $m = 7$—have more than one primitive root. The aim of this section is to prove a result that completely characterizes all moduli that possess at least one primitive root. We will also obtain some information about the number of primitive roots possessed by some special types of moduli (see, for instance, page 73, problem 2).

Theorem 3.6 – The Primitive Root Theorem: The positive integers possessing primitive roots are precisely the numbers 1, 2, 2^2, p^e, and $2p^e$, where p runs through all odd primes and e runs through all positive integers.

Remarks:

1. The argument we will use in proving this theorem involves the individual study of a number of different cases, and I will deliberately ask you to fill in quite a few of the details (see page 73, problem 1(a)–(g)).
2. In the case of odd prime modulus p, I will allow myself the luxury of invoking **Lagrange's Theorem.** This result will be stated and proved in Section 4.1 (page 83), but it will just be quoted at this stage in order to streamline the discussion.

   On to the proof!

Proof: The moduli 1, 2, and 2^2 can be checked by hand (please!), and we start by considering the case where the modulus $m = p$, with p being any arbitrary but fixed odd prime. Our job is to establish that there exists at least one integer, call it g, for which $\text{ord}_p g = \phi(p) = p - 1$. We may confine our search to the set $\{1, 2, 3, \ldots, p - 1\}$ (why?). Regardless of the choice of the integer b from this set, we have $\text{ord}_p b = d$, where d is necessarily some divisor of $p - 1$, according to property 1 of the order symbol (see page 67). The positive integer d is, of course, a function of the preassigned integer b. We need to show that $d = p - 1$ for at least one choice of b. For fixed $d|(p - 1)$, we denote by $\psi(d)$ the total number of integers b in the range $\{1, 2, 3, \ldots, p - 1\}$ for which $\text{ord}_p b = d$. Since $b^d \equiv 1 \pmod p$, it follows that $(b^2)^d \equiv 1 \pmod p$, $(b^3)^d \equiv 1 \pmod p$, $\ldots$, $(b^{d-1})^d \equiv 1 \pmod p$, and finally $(b^d)^d \equiv 1^d \equiv 1 \pmod p$; so the dth-degree polynomial congruence $X^d \equiv 1 \pmod p$ has at least d solutions—namely, $1, b, b^2, \ldots, b^{d-1}$ (note that these d integers are pairwise incongruent modulo

p—why?). It is precisely here that Lagrange's Theorem comes to our aid: it tells us that $\{1, b, b^2, \ldots, b^{d-1}\}$ is the entire solution set for this congruence.[1] Furthermore, we can conclude (using property 4 of the order symbol, given on page 67) that $\psi(d) = \phi(d)$, provided that there are any elements b of order d at all. We now have the following facts:

1. For every $d|\phi(p) = p - 1$, either $\psi(d) = 0$ or $\psi(d) = \phi(d)$.
2. $\sum_{d|p-1} \phi(d) \geqslant \sum_{d|p-1} \psi(d) = \phi(p)$.

Since $\sum_{d|p-1} \phi(d) = \phi(p) = p - 1$ (see page 70, problem 25), it follows that $\psi(d) = \phi(d)$ for every divisor d of $p - 1 = \phi(p)$. Specializing to the case of current interest, namely $d = p - 1$, tells us that $\psi(p - 1) = \phi(p - 1)$; and since $\phi(p - 1) > 0$, it follows that there is at least one element, call it g, satisfying $1 \leqslant g \leqslant p - 1$ with $\text{ord}_p g = p - 1$.

Now that we know that every odd prime p has at least one primitive root, we can use this information to help us prove that p^e has primitive roots, regardless of the choice of the positive integer e. Specifically, we will show that, if g is any primitive root of the odd prime p, there exists an integer x such that $g' = g + px$ is a primitive root of p^e for all positive integers e. Since $g^{p-1} \equiv 1 \pmod p$, we may write $g^{p-1} = 1 + py$ for some integer y. Now, by the Binomial Theorem,[2]

$$\begin{aligned}(g')^{p-1} &= (g + px)^{p-1} \\ &= g^{p-1} + \binom{p-1}{1} g^{p-2}px + p^2 \sum_{k=2}^{p-1} \binom{p-1}{k} g^{p-k-1}p^{k-2}x^k \\ &= 1 + py + p(p-1)g^{p-2}x + p^2 \sum_{k=2}^{p-1} \binom{p-1}{k} g^{p-k-1}p^{k-2}x^k\end{aligned}$$

and so we may define an integer z via the equation $(g')^{p-1} = 1 + pz$, in conjunction with the congruence $z \equiv y + (p - 1)g^{p-2}x \pmod p$. (Please verify!) The linear (in x) congruence $(p - 1)g^{p-2}x \equiv z - y \pmod p$ is solvable regardless of the choice of z (why?). I now claim that, if we choose—as we may—the integer z to be relatively prime to p, then any resultant $g' = g + px$ will be a primitive root modulo p^e for all positive integers e. To establish the claim, we notice first that $\phi(p) = p - 1 = \text{ord}_p g' | \text{ord}_{p^e}(g') | p^{e-1}(p - 1)$, using the fact that g' is a primitive root modulo p (why is g' a primitive root modulo p?). If we write $\text{ord}_{p^e}(g') = p^k(p - 1)$, where $k \leqslant e - 1$, then $(g')^{\text{ord}_{p^e}(g')} = ((g')^{p-1})^p = (1 + pz)^p = 1 + p^{k+1}z_k$ for some integer z_k, which is relatively prime to p (why?). Hence $p^{k+1} \| ((g')^{\text{ord}_{p^e}(g')} - 1)$, while $p^e | ((g')^{\text{ord}_{p^e}(g')} - 1)$; so $e \leqslant k + 1$ (i.e., $e - 1 \leqslant k$), and thus finally $k = e - 1$, as desired.

[1] in the sense of Definition 3.3.1 on page 55.

[2] See 6., Appendix I, page 108.

Armed with the knowledge that every positive integral power (e) of every odd prime (p) possesses at least one primitive root (g'), it is easy for us to establish that every modulus of the form $2p^e$ has the same property (see problem 1(f) on page 73).

To complete the proof of the Primitive Root Theorem, it is enough to show that no number of the form 2^e, with $e > 2$, has a primitive root and, finally, that no number of the form $n = n_1n_2$, with $n_1 > 2$, $n_2 > 2$ and $(n_1, n_2) = 1$, has a primitive root. It is not difficult to prove that $b^{2^{e-2}} \equiv 1 \pmod{2^e}$, regardless of the odd integer b (see problem 1(g) on page 73) and regardless of the integer $e > 2$, and thus the first of our two remaining jobs is done (why?). So consider $n = n_1n_2$, with $n_1 > 2$, $n_2 > 2$, and $(n_1, n_2) = 1$. Then $b^{\phi(n)/2} = (b^{\phi(n_1)})^{\phi(n_2)/2} \equiv 1 \pmod{n_1}$ and, upon interchanging the roles of n_1 and n_2, $b^{\phi(n)/2} \equiv (b^{\phi(n_2)})^{\phi(n_1)/2} \equiv 1 \pmod{n_2}$; and hence, finally, $b^{\phi(n)/2} \equiv 1 \pmod{n}$. The proof is complete.

PROBLEMS

Solutions (or hints at solutions) to the problems marked with an asterisk (*) are given on page 128.

*1. Fill in all the "gaps" in the proof of the Primitive Root Theorem by responding to the following requests.
   (a) Verify that the moduli 1, 2, and 2^2 possess primitive roots.
   *(b) If the odd prime p has a primitive root, it will have a primitive root belonging to the set $\{1, 2, 3, \ldots, p - 1\}$. Why?
   (c) Prove that there exists an integer z satisfying the conditions $(g')^{p-1} = 1 + pz$ and $z \equiv y + (p - 1)g^{p-2}x \pmod{p}$, as was claimed on page 72. In addition, show that we may arrange for $(z, p) = 1$.
   (d) Prove that the congruence $(p - 1)g^{p-2}x \equiv z - y \pmod{p}$ is solvable (for x), regardless of the choice of the integer z (see page 72).
   *(e) Why is g' a primitive root of p? (see page 72).
   (f) Prove that every modulus of the form $2p^e$ (notation as throughout Section 3.6) has a primitive root. [HINT: If g' is a primitive root of p^e, then look at the odd element of the pair $\{g', g' + p^e\}$.]
   *(g) Prove that $b^{2^{e-2}} \equiv 1 \pmod{2^e}$, regardless of the odd integer b and regardless of the integer $e > 2$, thus establishing that no modulus of this form has a primitive root.

*2. Is it correct to claim that the odd prime p has exactly $\phi(\phi(p))$ primitive roots? Give a reason for your answer. [HINT: Look back at the appropriate portion of the proof of the Primitive Root Theorem.]

3. Prove that, if $p > 3$ is a prime, then the product of all the primitive roots of p is congruent to 1 modulo p. [HINT: Represent all the primitive roots of the prime p in terms of a single primitive root.]

4.(a) If g is a primitive root of the prime p and if $(k, p - 1) = 1$, does it follow that g^k is also a primitive root modulo p? Prove that your answer is correct.

*(b) How many primitive roots does a prime have? Prove that your answer is correct. Verify your formula in the cases of $p = 5, 7$, and 11, by actually finding all of the primitive roots for these particular primes.

*5. Assume that $q = 4^n + 1$ for some positive integer n. Further, assume that $3^{(q-1)/2} \equiv -1 \pmod{q}$. Must q be a prime? Proof? (See page 103, problem 7.)

3.7 nth Power Residues, Indices, and Euler's Criterion

In this final section of Chapter 3, we provide a brief introduction to yet another variation on the general $a^b \equiv c \pmod{d}$ theme—namely, the concept of nth power residue.

Definition 3.7.1: Assume that the integer b is relatively prime to the modulus $m > 1$. We say that b is an **nth power residue modulo m** if and only if the nth degree polynomial congruence[1] $X^n \equiv b \pmod{m}$ is solvable.

Example 3.7.1: The quadratic ($n = 2$) residues modulo 5 are all the integers b such that $b \equiv 1 \pmod{5}$ or $b \equiv 4 \pmod{5}$. A truly brute force way of discovering this fact is to consider the second-degree congruence $X^2 \equiv b \pmod{5}$ for all integral b. However, if $b \equiv c \pmod{5}$, then the quadratic character of b and c is the same (why?); i.e., either b and c are both quadratic residues modulo 5 or both b and c fail to be quadratic residues modulo 5, in which case we refer to them as quadratic nonresidues modulo 5. It is therefore enough to check all elements of any arbitrary but fixed reduced residue system modulo 5. For instance, $1^2 \equiv 1 \pmod{5}$, $2^2 \equiv 4 \pmod{5}$, $3^2 \equiv 4 \pmod{5}$, and $4^2 \equiv 1 \pmod{5}$. This tells us that $\{1, 4\}$ is a complete set of pairwise incongruent quadratic residues modulo 5, and hence that $\{2, 3\}$ is a complete set of pairwise incongruent quadratic nonresidues modulo 5.

[1] See Definition 4.1.2, page 82.

We might notice a number of things about this (admittedly very simple) first example. For starters, the number of quadratic residues is equal to the number of quadratic nonresidues.[1] Then, too, each of the second-degree (quadratic) congruences $X^2 \equiv 1 \pmod 5$ and $X^2 \equiv 4 \pmod 5$ has exactly two solutions—the maximum possible number according to Lagrange's Theorem. It is reasonable to ask to what extent this type of behavior generalizes.

Example 3.7.2: The number b^n automatically serves as an nth power residue modulo m, provided that $(b^n, m) = 1$. For instance, since 1 and 4 are perfect squares, they are quadratic residues to every modulus m satisfying $(1, m) = (4, m) = 1$—i.e., to every odd modulus. In particular, 1 and 4 are quadratic residues modulo 5, as we already saw in Example 3.7.1.

Example 3.7.3: If we are looking for quadratic residues modulo 8, we must confine ourselves to odd values of the integer b. As in Example 3.7.1, it is enough to look through the elements of any arbitrary but fixed reduced residue system modulo 8, if we are looking for quadratic residues that are pairwise incongruent modulo 8. If we use the set $\{1, 3, 5, 7\}$ as our choice of reduced residue system modulo 8, we find that $X^2 \equiv 1 \pmod 8$ is solvable but that none of $X^2 \equiv 3 \pmod 8$, $X^2 \equiv 5 \pmod 8$, and $X^2 \equiv 7 \pmod 8$ is solvable. Hence 1 is a quadratic residue modulo 8, while 3, 5, and 7 are quadratic nonresidues modulo 8. Notice that in this case, unlike the situation in Example 3.7.1, the number of quadratic residues modulo 8 is not the same as the number of quadratic nonresidues modulo 8. Moreover, the second-degree congruence $X^2 \equiv 1 \pmod 8$ has four solutions. This fact, in conjunction with Lagrange's Theorem, provides a somewhat bizarre way of proving that the number 8 isn't a prime!

We shall study quadratic residues of a prime modulus systematically in Chapter 5, when we learn about the Law of Quadratic Reciprocity. For the time being, we content ourselves with stating and proving the following criterion related to nth power residues.

Theorem 3.7.1 – Euler's Criterion: Let m be a positive integer having a primitive root, and let $(b, m) = 1$. Then b is an nth-power residue modulo m if and only if $b^{\phi(m)/(n,\phi(m))} \equiv 1 \pmod m$.

This result can be established quite readily by arguing via indices. We thus prepare the way for the proof by defining the term *index* and establishing a couple of relevant properties of indices.

[1]Counted, as usual, according to Definition 3.3.1 on page 55.

Definition 3.7.1: If $m > 1$ is an integer that has a primitive root and if $(b, m) = 1$, then we define the **index** of b via the congruence $g^{\text{ind } b} \equiv b \pmod{m}$. Here, g is any arbitrary but fixed primitive root of the modulus m.

Remarks:

1. Since g is a primitive root of m, the $\phi(m)$ elements of the set $\{g, g^2, g^3, \ldots, g^{\phi(m)}\}$ form a reduced residue system modulo m. Hence the exponential congruence $g^X \equiv b \pmod{m}$ is solvable. (Please verify these statements.)
2. The index of b modulo m, ind b, depends not only on b and m, but also on the particular primitive root g (see page 77, problem 4 for a quantitative result).
3. Since $g^{\phi(m)} \equiv 1 \pmod{m}$ by the Euler-Fermat Theorem, ind b is only determined up to (integral) multiples of $\phi(m)$; i.e., $g^x \equiv b \pmod{m}$ if and only if $x \equiv \text{ind } b \pmod{\phi(m)}$.

We are now in a position to prove the above result.

Proof of Euler's Criterion: The congruence $X^n \equiv b \pmod{m}$ is solvable if and only if the congruence $(g^{\text{ind } X})^n \equiv b \pmod{m} \equiv g^{\text{ind } b} \pmod{m}$ is solvable, and this requirement in turn is equivalent to the solvability of $n \text{ ind } X \equiv \text{ind } b \pmod{\phi(m)}$. Now $n \text{ ind } X \equiv \text{ind } b \pmod{\phi(m)}$ is solvable if and only if $(n, \phi(m)) | \text{ind } b$. This divisibility relation may be recast as $\text{ind } b \equiv 0 \pmod{(n, \phi(m))}$ or $\{\phi(m)/(n, \phi(m))\} \text{ ind } b \equiv 0 \pmod{\phi(m)}$, which, finally, is equivalent to $b^{\phi(m)/(n,\phi(m))} = 1 \pmod{m}$. Q.E.D.

PROBLEMS

Solutions (or hints at solutions) to the problems marked with an asterisk (*) are given on page 128.

*1. For each modulus m, let $R = \{1 = r_1 < r_2 < \ldots < r_{\phi(m)} = m - 1\}$ be the "standard" reduced residue system modulo m. Determine precisely which elements of R are quadratic residues modulo m for $m = 7, 11, 13, 17; 6, 10, 12, 14$.

*2. Find all the cubic residues of the modulus $m = 9$ that are in the interval $[1, 8]$.

3. Assume that the prime modulus p satisfies $p > 15$ but is otherwise completely arbitrary. Prove that there must exist two consecutive quadratic residues—call them b and $b + 1$—with respect to the modulus p, with $b \leqslant 9$. [Hints: (1) 1 is always a quadratic residue modulo p, so assume that 2

is a quadratic nonresidue; otherwise, we may take $b = 1$. (2) $4 \equiv 2^2 \pmod{p}$. (3) Is the product of two quadratic residues modulo m always a quadratic residue modulo m? How about quadratic nonresidues?]

4. **"Change-of-Base Formula."** Prove that, if g and h are any two primitive roots of the prime p, then $\text{ind}_h b \equiv \text{ind}_g b \cdot \text{ind}_h g \pmod{\phi(p)}$.

*5. Prove that b is a quadratic residue modulo the odd prime p if and only if $b^{(p-1)/2} \equiv 1 \pmod{p}$. [HINT (strong!): Specialize Euler's Criterion appropriately.]

*6. Use the criterion of problem 5 to answer the following questions.
   *(a) Is 6 a quadratic residue of 7?
   *(b) Is 43 a quadratic nonresidue of 83?

*7. Assume that p is an odd prime and that $p \nmid b$. Use Fermat's Little Theorem to conclude that $b^{(p-1)/2} \equiv \pm 1 \pmod{p}$.

*8. *(a) Compute the index of each element of the "standard" reduced residue system modulo 9 (in the sense of page 76, problem 1) if $g = 2$.
   (b) Assume that $m = 18$ and that $g = 5$. Compute the index of each element of the "standard" reduced residue system modulo m.

9. Find a complete set of pairwise incongruent (modulo 10) quartic[1] residues modulo 10. Compare page 43, problem 3(b).

*10. Assume that p is any odd prime and that $p \nmid b$. Evaluate $|(b^{(p-1)/2} - 1) + (b^{(p-1)/2} + 1)|_p$ in two different ways.

11. Prove the following:
   (a) $\text{ord}_m b = \phi(m)/(\text{ind}_m b, \phi(m))$.
   (b) $b^x \equiv c \pmod{m}$ is solvable if and only if $\text{ord}_m c | \text{ord}_m b$, if and only if $(\text{ind}_m b, \phi(m)) | (\text{ind}_m c, \phi(m))$.

NOTES TO CHAPTER 3

1. The set $\{X \equiv 0 \pmod 2, X \equiv 0 \pmod 3, X \equiv 1 \pmod 4, X \equiv 5 \pmod 6, X \equiv 7 \pmod{12}\}$ of linear congruences from problem 9 on page 44 is an example of a **covering set of congruences.** See, for instance, page 57 of Harold Davenport's *The Higher Arithmetic* (a British synonym for "the theory of numbers"). See also page 77 of LeVeque's *Fundamentals of Number Theory.*

2. The result of problem 12 on page 44, in conjunction with Wilson's Theorem, provides us with yet another characterization of primes.

3. Fermat's Little Theorem (Corollary 3.2.1 on page 50)—a very useful result in its own right—owes its name to the contrast between itself and

[1] i.e., fourth-degree.

Fermat's GREAT (Theorem): regardless of the choice of the integer $n > 2$, the equation $x^n + y^n = z^n$ has no solutions in positive integers x, y, z. The word *Theorem* is in parentheses because the statement is still in question—no proof having been given to date. Gerd Faltings' recent (1983) extremely deep work in establishing Mordell's Conjecture[1] allows one to conclude that there exist only finitely many rational solutions of the Fermat equation for any particular choice of $n > 2$. Other important recent work on Fermat's Last (Theorem) has been done by Adleman and Heath-Brown, Filaseta, Granville, Ribet, and Serre.

4. The situation with respect to primitive Pythagorean triples (page 52, problem 10) is very satisfactory, in the following sense: (x, y, z) is a primitive Pythagorean triple if and only if $x = r^2 - s^2$, $y = 2rs$, and $z = r^2 + s^2$ for some pair of integers r and s satisfying the constraints that $r > s > 0$, $(r, s) = 1$, and r and s must be of opposite parity (i.e., one of the pair r and s is odd, and the other is even). Thus, the solution set of the Diophantine equation $x^2 + y^2 = z^2$ is completely known. We say that the curve represented by this equation can be **rationally parameterized.** See Niven and Zuckerman's *An Introduction to the Theory of Numbers,* pages 136–39, for a proof of the above theorem. More advanced references include *Studies in Number Theory,* Volume 6, MAA Studies in Mathematics, edited by W. J. LeVeque, and the fascinating recent research-expository paper by Barry Mazur entitled "Arithmetic on Curves," in the April 1986 issue of the *Bulletin of the American Mathematical Society,* pages 207–59.
5. The Chinese Remainder Theorem is strongly related to something called the Approximation Theorem in valuation theory. George Bachman's *Introduction to Valuation Theory and* p-*adic Numbers* is a good source; see specifically Theorem 3.3 (the Approximation Theorem), pages 21–22, and Exercise 9, page 23.
6. **Multiplicativity.** This term has already been encountered several times in the text, in a variety of contexts (page 23, problem 8(d), in connection with the number of divisors function τ; page 28, problem 8, in connection with the norm of elements of the set S; page 30, Equation (1), in connection with p-adic valuations; and most recently, in the remark on page 64, in connection with the Euler ϕ-function, and in Theorem 3.5.1 on page 64). The situation is complicated by the fact that there are two meanings for the term *multiplicative,* both of which are very important and have been adopted in standard terminology:
   (a) If one is dealing with an arithmetic (number-theoretic) function f—

[1]For which he received a Fields' Medal on August 3, 1986, during the opening ceremonies of the International Congress of Mathematicians, held at U.C. Berkeley.

such as τ or ϕ—then *multiplicative* means that $f(mn) = f(m)f(n)$ provided that the arguments m and n are relatively prime.

(b) If one is dealing with a norm or p-adic valuation function f, then *multiplicative* means that $f(mn) = f(m)f(n)$ for all arguments m and n in the domain of f.

The second meaning above is of extreme importance in abstract algebra, because it is related to the concepts of isomorphism (see Appendix II, page 114) and, more generally, homomorphism.

7. In Section 3.5 the concept of the order of an element relative to a modulus and the concept of a primitive root of a modulus were introduced. From the point of view of abstract algebra, $\text{ord}_m b$ is the order of the element b in the group $\{R; \cdot\}$ of problem 1(b) of Appendix II (page 117), while the existence of a primitive root of a modulus corresponds to the cyclic nature of the group $\{R; \cdot\}$.

8. D. H. Lehmer conjectured in 1932 that, if $\phi(n)|(n - 1)$, then n must be a prime. This conjecture is still outstanding. See also LeVeque's *Fundamentals of Number Theory.*

9. **Artin's Conjecture for Primitive Roots.** In 1927, Artin conjectured that any integer k, other than -1 or a perfect square, is a primitive root for infinitely many primes. Gupta and Ram Murty (1984) showed that there are infinitely many k for which Artin's Conjecture is true. Heath-Brown (1986) improved the method used by Gupta and Ram Murty and obtained yet stronger results. In particular, he proved that there are at most two primes for which Artin's Conjecture fails.

10. The concept of *index* (see Definition 3.7.1 on page 76) is sort of a discrete or number-theoretic logarithm: $\text{ind}\,(bc) \equiv \text{ind}\, b + \text{ind}\, c \pmod{\phi(m)}$ and $\text{ind}\, b^n \equiv n \cdot \text{ind}\, b \pmod{\phi(m)}$ can easily be deduced from the defining relation. See also problem 4 on page 77, for the Change-of-Base Formula analogous to that for logarithms.

4 Polynomial Congruences in One Variable

INTRODUCTION

In Section 3.3, we learned something about individual linear (first-degree) and quadratic (second-degree) congruences. In Section 3.4, we studied finite systems of simultaneous linear congruences—the Chinese Remainder Theorem. Finally, in Section 3.6, we became familiar with the statement (at least) of Lagrange's Theorem—a result that applies to polynomial congruences of arbitrary degree but prime modulus.

My goal in this chapter is to outline the general situation $f(X) \equiv 0 \pmod{m}$, where $f(X)$ is a polynomial of arbitrary degree but having integral coefficients and were $m > 1$ is any modulus.

Let $Z[X]$ denote the set of all polynomials in the indeterminate X with integral coefficients. Suppose that $f(X)$ is an element of $Z[X]$ and that the modulus m has canonical decomposition $m = p_1^{e_1} \cdot p_2^{e_2} \cdot \ldots \cdot p_k^{e_k}$. Then the problem of finding an integer x for which $f(x) \equiv 0 \pmod{m}$ is equivalent to the problem of finding an integer x for which the simultaneous congruences $f(x) \equiv 0 \pmod{p_1^{e_1}}$, $f(x) \equiv 0 \pmod{p_2^{e_2}}, \ldots, f(x) \equiv 0 \pmod{p_k^{e_k}}$ are satisfied. So we are led, in this natural canonical decomposition manner, to study polynomial congruences to a prime power modulus; i.e., $f(X) \equiv 0 \pmod{p^e}$, where p is any prime and e is any positive integer. Any solution, say x, of a congruence of the type $f(X) \equiv 0 \pmod{p^e}$ is, of necessity, a solution of each of the congruences $f(X) \equiv 0 \pmod{p^{e-1}}$, $f(X) \equiv 0 \pmod{p^{e-2}}, \ldots, f(X) \equiv 0 \pmod{p^2}$, and finally $f(X) \equiv 0 \pmod{p}$ (the converse of this statement is—as we shall see—highly false, in general). We have thus, once again, been led from a congruence involving a more complicated modulus—in this case, a prime power—to a corresponding congruence with respect to a simpler modulus—in this case, a prime. We can go no further in simplifying the multiplicative structure of the modulus because of what it means to be prime, so we are at ground level now. The plan of this chapter is to start at ground level, to learn something about polynomial congruences to a prime modulus, and then to crawl back up the ladder through the prime power modulus stage to the full generality of the original modulus m. It is important to realize how natural a tool **the *p*-adic valuation approach** is in analyzing how one "lifts" from the prime modulus p to the prime power modulus p^e.

4.1 Polynomial Congruences with Respect to a Prime Modulus; i.e., $f(X) \equiv 0 \pmod{p}$

I feel morally obligated to prove Lagrange's Theorem first, since I have already made use of it (at pages 72 and 75). In order to understand even the statement of Lagrange's Theorem, we need to have two definitions clearly in mind—namely:

1. The definition of **the number of solutions** of a polynomial congruence.
2. The definition of the **degree** of a polynomial congruence.

The first of these two definitions was introduced—at least for the special cases of linear and quadratic congruences—in Section 3.3 (see page 55, Definition 3.3.1). We now restate it in a slightly different form for polynomial congruences in general.

Definition 4.1.1: Let $\{x_1, x_2, \ldots, x_m\}$ be any arbitrary but fixed complete residue system modulo m. The **number of solutions** of the polynomial congruence $f(X) \equiv 0 \pmod{m}$ is the number of elements x_i such that $f(x_i) \equiv 0 \pmod{m}$.

The second definition was not explicitly introduced previously.

Definition 4.1.2: Let $f(X) = b_nX^n + b_{n-1}X^{n-1} + \ldots + b_1X^1 + b_0$ be an element of $Z[X]$ of degree n; i.e., $b_n \neq 0$. The **degree** of the corresponding polynomial congruence $f(X) \equiv 0 \pmod{m}$ is defined to be k, provided that $b_k \not\equiv 0 \pmod{m}$ but $b_n \equiv b_{n-1} \equiv \ldots \equiv b_{k+1} \equiv 0 \pmod{m}$. If no such integer k exists, we don't attach a degree to the corresponding congruence.

Remarks:

1. It is important to bear in mind that the congruential degree (if it exists) is invariably less than or equal to the degree of the corresponding polynomial.

2. The congruential degree, k, is a function of the modulus m; i.e., $k = k(m)$. Hence, a single polynomial will, in general, lead to polynomial congruences of differing degrees, depending on the particular moduli chosen. (See page 86, problems 1 and 3, for instance.)

We are now in a position to state and prove Lagrange's Theorem.

Theorem 4.1.1 – Lagrange's Theorem: The polynomial congruence $f(X) \equiv 0 \pmod{p}$ has no more than n solutions, provided $f(X)$ is a polynomial of degree n with leading coefficient not divisible by the prime modulus p.

Proof: Our proof proceeds by induction on n. The case $n = 0$ is trivial (why?). Although, strictly speaking, we don't have to verify the case $n = 1$, it follows from what we previously learned about linear congruences (see page 55). We take as our inductive hypothesis the validity of the statement for polynomial congruences of degree $n - 1$ modulo p. Now $f(X) - f(b) = (X - b)g(X)$, where $g(X)$ is a polynomial of degree $n - 1$, with leading coefficient not divisible by p (why?). This statement is valid regardless of the choice of the integer b. An integer b is a solution of $f(X) \equiv 0 \pmod{p}$ if and only if it is a solution of $(X - b)g(X) \equiv 0 \pmod{p}$; and by the inductive hypothesis, this latter congruence has no more than n solutions. Q.E.D.

Remark: It might be interesting to compare/contrast Lagrange's Theorem with the so-called Fundamental Theorem of Algebra: If $f(X)$ is an element of $C[X]$, C being the field of complex numbers, then the number of solutions of the polynomial equation $f(X) = 0$ agrees precisely with the degree of the polynomial $f(X)$ (provided that a solution of multiplicity 2, for example, is counted as 2 solutions).

Example 4.1.1: We have already seen (Example 3.7.3 on page 75) that the congruence $X^2 \equiv 1 \pmod{8}$ has too many solutions for 8 to be a prime. Lagrange's Theorem guarantees that the congruence $3X \equiv 4 \pmod{5}$ has no more than one solution and, in fact, that $\{\underline{3}\}$ can be regarded as a complete set of solutions for this congruence. We know that the congruence $X^3 - X + 7 \equiv 0 \pmod{3}$ can have no more than three solutions. However, since $b^p \equiv b \pmod{p}$ for all integers b whenever p is a prime (why?), this cubic congruence has no solutions at all.

We can further exploit the idea behind the last example. The congruence $X^p - X \equiv 0 \pmod{p}$, p being any prime, has a maximum of p solutions—for at least two different reasons. One reason can be found in Lagrange's Theorem. In this case, we are focusing our attention on the following facts: the polynomial $X^p - X$ has degree p; the polynomial $X^p - X$ has leading coefficient not divisible by the modulus of the congruence; and the modulus of the congruence is a prime (the fact that the modulus is the particular prime p is irrelevant at this point). A second and much easier way of realizing that our congruence has no more than p solutions is simply to keep in mind that p is the magnitude of the modulus, together with Definition 4.1.1 on page 82. (Here it is irrelevant that p is a prime.) As a matter of fact, there is a third way of seeing that the congruence $X^p - X \equiv 0 \pmod{p}$, p being any prime, has a maximum of p solutions, and it is definitely preferable to the two ways already mentioned: simply invoke Fermat's Little Theorem (see page 50), and conclude immediately that $X^p - X \equiv 0 \pmod{p}$, p being any prime, has precisely p solutions.

The existence of this **pth Fermat polynomial** $f_p(X) = X^p - X$ allows us to show quite easily that, when solving polynomial congruences modulo the prime p, we don't have to tolerate any degree in excess of p; or to say it more positively, we can assume without loss of generality that the degree of $f(X)$ is strictly less than p if we want to solve $f(X) \equiv 0 \pmod{p}$. For suppose that the polynomial $f(X)$ initially has degree greater than or equal to p. Upon dividing $f(X)$ by the pth Fermat polynomial $f_p(X) = X^p - X$, we get an equation of the form $f(X) = (X^p - X)g(X) + r(X)$, where $g(X)$ and $r(X)$ are elements of $Z[X]$ and where either the remainder $r(X)$ is identically zero (i.e., the zero polynomial) or the degree of $r(X)$ is strictly less than p. This "long division" of polynomials—the Division Algorithm, more formally—doesn't lead to rational, nonintegral coefficients in either $q(X)$ or $r(X)$ because the leading coefficient of the divisor polynomial, $X^p - X$, is 1. The equation $f(X) = (X^p - X)q(X) + r(X)$ makes it clear that the solution set of the congruence $f(X) \equiv 0 \pmod{p}$ is precisely the same as the solution set of the congruence $r(X) \equiv 0 \pmod{p}$. In the two special cases where either $r(X)$ is identically zero or every coefficient of $r(X)$ is divisible by p, then $f(X) \equiv 0 \pmod{p}$ has the maximum possible number of solutions—namely, p. In all other cases, the original problem $f(X) \equiv 0 \pmod{p}$ has been replaced by the equivalent problem $r(X) \equiv 0 \pmod{p}$, where the degree of $r(X)$ is strictly less than p.

We finish this brief introduction to the study of polynomial congruences to a prime modulus by establishing a criterion for deciding precisely when the upper bound on the number of solutions of $f(X) \equiv 0 \pmod{p}$ given by Lagrange's Theorem is achieved.

Theorem 4.1.2: The polynomial congruence $f(X) \equiv 0 \pmod{p}$—with p a prime, with $f(X)$ of degree n, and with leading coefficient 1—has n solutions if and only if there exist $q(X)$, $r(X)$ in $Z[X]$ for which $X^p - X = f(X)q(X) + pr(X)$, where either $r(X)$ is identically zero or the degree of $r(X)$ is less than n.

Proof: If our congruence has n solutions, then necessarily $n \leqslant p$ (remember Definition 4.1.1, page 82). Application of the Division Algorithm to $X^p - X$ and $f(X)$ leads to $X^p - X = f(X)q(X) + R(X)$, where either $R(X)$ is identically zero or the degree of $R(X)$ is strictly less than n. This equation tells us that every solution of $f(X) \equiv 0 \pmod{p}$ is a solution of $R(X) \equiv 0 \pmod{p}$. If $R(X)$ isn't identically zero, then $R(X)$ is a polynomial of degree less than n, while the congruence $R(X) \equiv 0 \pmod{p}$ has n solutions. These two constraints require that every coefficient of $R(X)$ be divisible by p, so that $R(X) = pr(X)$ for suitable $r(X)$ satisfying the conditions of the theorem.

Now assume, conversely, that $X^p - X = f(X)q(X) + pr(X)$, as in the statement of the theorem. It follows that the congruence $f(X)q(X) \equiv 0 \pmod{p}$ has p solutions. From the equation and the standing assumptions on $f(X)$, it follows (making two applications of Lagrange's Theorem) that the congruences $f(X) \equiv 0 \pmod{p}$ and $q(X) \equiv 0 \pmod{p}$ have no more than n—respectively, $(p - n)$—solutions. Since $f(X)q(X) \equiv 0 \pmod{p}$ forces either $f(X) \equiv 0 \pmod{p}$ or $q(X) \equiv 0 \pmod{p}$, it follows that each of the congruences $f(X) \equiv 0 \pmod{p}$, $q(X) \equiv 0 \pmod{p}$ must have its full complement of solutions. In particular, $f(X) \equiv 0 \pmod{p}$ must have p solutions.

Sharper results can be had if one is willing to use more abstract algebra. See, for instance, Theorem 3.19 on page 64 of LeVeque's *Fundamentals of Number Theory,* where the context is taken to be $F_p[X]$—the ring of polynomials with coefficients from the finite field of p elements, F_p.

Example 4.1.2–Wilson's Theorem revisited: One of the prettiest applications of the preceding circle of ideas is, I think, the following alternate way of proving Wilson's Theorem (see page 41). Fermat's Little Theorem tells us that $b^{p-1} \equiv 1 \pmod{p}$ for every integer b such that $(b, p) = 1$, p being any arbitrary but fixed prime. Lagrange's Theorem tells us that $X^{p-1} - 1 \equiv 0 \pmod{p}$ has no more than $p - 1$ solutions, but Fermat's Little Theorem tells us that this polynomial congruence has the full complement—namely, $(p - 1)$—of solutions, and hence gives stronger information than does Lagrange's Theorem. We may take a complete solution set of $X^{p-1} - 1 \equiv 0 \pmod{p}$ to be $\{1, 2, 3, \ldots, p - 1\}$, the most natural choice. Since $X^{p-1} - 1 \equiv 0 \pmod{p}$ has the maximum number of solutions that the law—and Lagrange—allows, we know from Theorem 4.1.2 on page 84 that $X^p - X = (X^{p-1} - 1)q(X) + pr(X)$ for suitable $q(X)$, $r(X)$ in $Z[X]$. In this case, $q(X) = X$, and $r(X)$ is identically zero. Now $X^{p-1} - 1 \equiv 0 \pmod{p}$ doesn't have zero as a solution, but, courtesy of Fermat, the polynomial congruence $X^{p-1} - 1 \equiv (X - 1) \cdot (X - 2) \cdot \ldots \cdot (X - (p - 1)) \pmod{p}$ has at least $1, 2, \ldots, p - 1$ as solutions. However, we now have a polynomial congruence $X^{p-1} - 1 - [(X - 1) \cdot (X - 2) \cdot \ldots \cdot (X - (p - 1))] \equiv 0 \pmod{p}$ of degree no more than $(p - 2)$ (why?), with at least $(p - 1)$ solutions; and so every coefficient of the polynomial must be divisible by p (why?). The constant term of this polynomial is $(-1) - [(-1) \cdot (-2) \cdot \ldots \cdot (-(p - 1))] = (-1) - (p - 1)!$ ($p = 2$ is, as usual, exceptional, so treat it separately), so $p | [(-1) - (p - 1)!]$; i.e., $(p - 1)! \equiv -1 \pmod{p}$, Wilson's Theorem!

We can squeeze a little more mileage out of this line of reasoning. By now, $X^{p-1} - 1 \equiv (X - 1) \cdot (X - 2) \cdot \ldots \cdot (X - (p - 1)) \pmod{p}$ is known to have p solutions (is Lagrange's Theorem being violated at this point?). To illustrate the sort of result that we can glean from this congruence, let us note that $1 + 2 + 3 + \ldots + (p - 1) \equiv 0 \pmod{p}$ is a result of comparing the coefficients of X^{p-2} on the left- and right-hand sides of our congruence. Of course, $1 + 2 + 3 + \ldots + (p - 1) = (p - 1)p/2$ is better information and is a standard result, so this isn't very exciting. However, comparison of the coefficients of X^{p-3} leads to $(1 \cdot 2) + (1 \cdot 3) + (2 \cdot 3) + \ldots + (p - 2) \cdot (p - 1) = \sum_{1 \leq i < j \leq p-1} ij \equiv 0 \pmod{p}$, which is a little more exotic.

PROBLEMS

Solutions (or hints at solutions) to the problems marked with an asterisk (*) are given on page 129.

*1. In each of the following situations, find the degree of the polynomial $f(X)$ in $Z[X]$, and find the degree of the corresponding polynomial congruences $f(X) \equiv 0 \pmod{p}$, for $p = 2, 3$, and 7:
   *(a) $f(X) = X^6 + X^5 + X^4 + X^3 + X^2 + X + 1$
   (b) $f(X) = 2X^4 + 6X^3 - 7$
   *(c) $f(X) = X - X^4 + 42X^{91}$

*2. Use Lagrange's Theorem to obtain an upper bound for the number of solutions for each of the nine polynomial congruences of the form $f(X) \equiv 0 \pmod{p}$ as p runs through the primes $p = 2, 3$, and 7 and as $f(X)$ runs through the polynomials listed in problem 1.

*3. Find a complete set of solutions for each of the nine polynomial congruences described in problem 2.

4.(a) Could you improve on Lagrange's Theorem (see page 83) by simply omitting the hypothesis that the modulus be prime? Proof?
   *(b) Is it correct to claim that the number of solutions of the polynomial congruence $f(X) \equiv 0 \pmod{p}$, p being a prime, is bounded above by $\min\{n, p\}$? Why? (Here, n is the degree of the polynomial $f(X)$.)

We make the following definition in connection with problems 5, 6, and 7.

Definition 4.1.3: A polynomial congruence $f(X) \equiv 0 \pmod{m}$ is said to be an **identical congruence** if and only if $f(z) \equiv 0 \pmod{m}$ for all integers z.

5.(a) If $f(X) = b_nX^n + \ldots + b_0$ and $g(X) = c_nX^n + \ldots + c_0$ are elements of $Z[X]$ that satisfy $b_k \equiv c_k \pmod{m}$ for all k such that $0 \leqslant k \leqslant n$, does it follow that $f(X) - g(X) \equiv 0 \pmod{m}$ is an identical congruence? Prove that your answer is correct.

*(b) Is the converse of the above implication valid? Proof?

6. Suppose that $f(X)$ is an element of $Z[X]$ of degree n. Suppose further that $p|f(b)$, for some prime p and for at least $(n + 1)$ consecutive integers b. Conclude that $p|f(z)$, for all integers z—i.e., that $f(X) \equiv 0 \pmod{p}$ is an identical congruence.

7. (Continuation/generalization of problem 6.) Suppose that $f(X)$ is an element of $Z[X]$. Suppose further that there are at least m consecutive integers—say $b, b + 1, \ldots, b + m - 1$—for which $m|f(b + k)$, $0 \leqslant k \leqslant m - 1$. Prove that $f(X) \equiv 0 \pmod{m}$ is an identical congruence. Find an example of an $f(X) = b_nX^n + b_{n-1}X^{n-1} + \ldots + b_1X + b_0$ in $Z[X]$ satisfying $(b_n, b_{n-1}, \ldots, b_1, b_0) = 1$ and an $m > 1$ for which the above situation occurs.

*8. In each of the following situations, replace the original congruence with a congruence of degree less than the given prime p, by making use of the technique outlined on page 84:

(a) $2X^{15} + X^{11} - X^8 + 5 \equiv 0 \pmod{7}$.

*(b) $-X^2 - X^{18} \equiv 0 \pmod{17}$.

*(c) $X^6 + X^5 + X^4 + X^3 + X^2 + X^1 + 1 \equiv 0 \pmod{3}$.

(d) $2X^3 + 5X^2 + 6X + 1 \equiv 0 \pmod{7}$.

*9. Is each assertion below true or false? Justify your answer.

*(a) $|b^p - 1|_p < 1$ if and only if $|b|_p \geqslant 1$, given that b is an integer and that p is a prime.

*(b) $|b^p - b|_p \leqslant 1/p$ if and only if $|b|_p < 1$, given that b is an integer and that p is a prime.

*(c) $|b^p - b|_p \leqslant 1/p$ if and only if $|b|_q \leqslant 1$ for all primes q, given that b is an integer and that p is prime.

10. Apply Theorem 4.1.2 (page 84) to each part of problem 8 to decide whether the number of solutions of the resulting congruence agrees with the degree of that congruence.

4.2 Polynomial Congruences with Respect to a Prime Power Modulus; i.e., $f(X) \equiv 0 \pmod{p^e}$

We now start trying to crawl back up the ladder, in the sense mentioned in the introduction to Chapter 4. Suppose we know a solution, say x_0, of the polynomial congruence $f(X) \equiv 0 \pmod{p}$ for some prime modulus p. Then the problems we set ourselves are these:

1. Determine whether the congruence $f(X) \equiv 0 \pmod{p^2}$ has any solutions that "emanate" from the given solution, x_0, of $f(X) \equiv 0 \pmod{p}$.
2. If there <u>are</u> solutions under the terms of problem 1, try to describe all such solutions as accurately as possible.

You will notice that problems 1 and 2 aren't precisely stated as yet; we'll become more precise as we go along. I claim that the <u>right</u> way to look at this situation is as <u>an application of Newton's method of successive approximations</u>, but <u>from a p-adic point of view.</u>

Let's start by assuming that $f(x_0) \equiv 0 \pmod{p^e}$ and go hunting for an integer x_1 for which $f(x_1) \equiv 0 \pmod{p^{e+1}}$. The congruence $f(x_0) \equiv 0 \pmod{p^e}$ is equivalent to the divisibility statement $p^e | f(x_0)$, which in turn is equivalent to the p-adic statement $|f(x_0)|_p \leqslant p^{-e}$. The point is that when e is large, p^{-e} is very small (using $|\cdot|$, the usual measuring stick), which means that x_0 is a very good <u>p-adic</u> approximation to a solution of the <u>equation</u> $f(X) = 0$. If $e = 1$, we may regard x_0 as a p-adic first approximation to a solution, say α, of the equation $f(X) = 0$.

Let's see if we can refine this p-adic first approximation in order to obtain one or more p-adic second approximations to α—i.e., solutions of the congruence $f(X) \equiv 0 \pmod{p^2}$. The single solution, x_0, of $f(X) \equiv 0 \pmod{p}$ fans out into precisely $p^1 = p^2/p^1$ candidates for solutions of $f(X) \equiv 0 \pmod{p^2}$—namely, $x_0 + pt$, for $t = 0, 1, 2, \ldots, (p - 1)$. (Why does each of these p integers have a chance of being a solution of $f(X) \equiv 0 \pmod{p^2}$, and why is this list of candidates exhaustive? See page 90, problem 3). The two problems we set ourselves at the beginning of this section now take the following form:

1. Determine whether there is a value of t for which $f(x_0 + pt) \equiv 0 \pmod{p^2}$.
2. If there is at least one such value of t, identify exactly how many such values of t there are.

The ideal tool for this job comes from calculus—the so-called *Taylor polynomial* of $f(x) = b_nx^n + b_{n-1}x^{n-1} + \ldots + b_1x^1 + b_0$, now thought of as a function of the real variable x about the point x_0. In our situation, the Taylor polynomial of $f(x_0 + pt)$ about x_0 is given by the formula $f(x_0 + pt) = f(x_0) + f^{(1)}(x_0)(pt) + f^{(2)}(x_0)(pt)^2/2! + \ldots + f^{(n)}(x_0)(pt)^n/n!$, where $f^{(k)}(x_0)$, with $1 \leqslant k \leqslant n$, denotes the kth derivative of f, evaluated at x_0. The typical term in this Taylor polynomial is $f^{(k)}(x_0)(pt)^k/k!$. The typical contribution—from $f(x) = b_nx^n + b_{n-1}x^{n-1} + \ldots + b_1x^1 + b_0$—to this typical term is what we get by differentiating b_ix^i a total of k times and evaluating the result at x_0: $i(i - 1) \cdot \ldots \cdot (i - k + 1)b_ix_0^{i-k}$ (here $i \geqslant k$ is to be understood). Hence,

$$
\begin{aligned}
f^{(k)}(x_0)(pt)^k/k! &= \sum_{i=0}^{n} i(i-1)\cdot \ldots \cdot (i-k+1)b_ix_0^{i-k}(pt)^k/k! \\
&= \sum_{i=0}^{n} \binom{i}{k} b_ix_0^{i-k}(pt)^k
\end{aligned}
$$

is an integer since the binomial coefficient[1]

$$\binom{i}{k} = i!/k!(i-k)! = i(i-1)\cdot\ .\ .\ .\ \cdot(i-k+1)/k!$$

is known to be an integer. We can now level this machinery at our two problems. Since all terms in the Taylor expansion of $f(x_0 + pt)$ about x_0—except $f(x_0)$ and $f^{(1)}(x_0)(pt)$—are divisible by p^2 (why?), $f(x_0 + pt) \equiv f(x_0) + f^{(1)}(x_0)tp \pmod{p^2}$. Thus $f(x_0 + pt) \equiv 0 \pmod{p^2}$ if and only if $f^{(1)}(x_0)pt \equiv -f(x_0) \pmod{p^2}$ or, equivalently, $f^{(1)}(x_0)t \equiv -f(x_0)/p \pmod{p}$ (since $f(x_0) \equiv 0 \pmod{p}$, $f(x_0)/p$ is an integer). This linear congruence in t is solvable if and only if $(f^{(1)}(x_0), p) \mid -f(x_0)/p$. We may as well assume that $f(x_0) \not\equiv 0 \pmod{p^2}$, because otherwise we are already in possession of a p-adic second approximation to a solution of $f(X) = 0$. (Please see, however, page 90, Example 4.2.2.) So our congruence has no solutions if $p \mid f^{(1)}(x_0)$, and it has precisely $1 = (f^{(1)}(x_0), p)$ solution if $p \nmid f^{(1)}(x_0)$. (Recall the answer on pages 54–55 to question (b) at the beginning of Section 3.3.) Thus we have an easy way to decide when our first p-adic approximation "dies without progeny" (to use Bill LeVeque's descriptive phrase) or produces a unique child—x_1, say—this child serving as a p-adic second approximation to a solution of the equation $f(X) = 0$. This second approximation can be computed through the formulas $x_1 = x_0 + pt$ and $f^{(1)}(x_0)t \equiv -f(x_0)/p \pmod{p}$ to give

$$x_1 \equiv x_0 - f(x_0)/f^{(1)}(x_0) \pmod{p^2} \tag{1}$$

See page 91, problem 5.

We have described in some detail how to pass from a first p-adic approximation x_0, satisfying $f(x_0) \equiv 0 \pmod{p}$, to the second p-adic approximation x_1, satisfying $f(x_1) \equiv 0 \pmod{p^2}$, provided that $p \nmid f^{(1)}(x_0)$. By iterating this procedure, we can successively crawl up the rungs of the ladder from the initial $f(X) \equiv 0 \pmod{p}$ state to any desired level $f(X) \equiv 0 \pmod{p^e}$. If x_p is the solution of $f(X) \equiv 0 \pmod{p^e}$, then $|f(x_p)|_p \leq p^{-e}$, and we have a quantitative grip on the degree of p-adic approximation being achieved.

Example 4.2.1: Consider the polynomial $f(X) = X^2 - 11$ and the prime modulus $p = 7$. We may take $x_0 = 2$ as a first 7-adic approximation to a solution of the polynomial equation $f(X) = X^2 - 11 = 0$. Since $7 \nmid f^{(1)}(x_0) = 2x_0 = 4$, $x_0 = 2$ produces the unique second 7-adic approximation $x_1 = x_0 + pt = 2 + (7 \cdot 2)$, upon solving $f^{(1)}(x_0)t \equiv -f(x_0)/p \pmod{p}$—i.e., $4t \equiv -(-7)/7 \pmod{7}$—subject to $0 \leq t \leq (p-1) = 6$. The third 7-adic approximation can be obtained by solving $f^{(1)}(x_1)t \equiv -f(x_1)/p^2 \pmod{p}$—i.e., $32t \equiv -5 \pmod{7}$—subject to $0 \leq t \leq 6$, to obtain $t = 4$; and thus $x_2 = 2 + (2 \cdot 7) + (4 \cdot 7^2)$.

[1] See, for instance, Appendix I, page 108.

Since $7 \nmid f^{(1)}(x_0)$, $7 \nmid f^{(1)}(x_1)$, . . . , this process can be continued to obtain any degree of 7-adic approximation we desire (see also page 91, problem 8).

Example 4.2.2: "Theory is nice, but experience has a way of subverting it." A situation that looks much like Example 4.2.1 but exhibits quite different behavior can be obtained by taking $f(X) = X^2 + X + 7$ and $p = 3$. If we take $x_0 = x_1 = 1$, then $3 | f^{(1)}(1)$, so that $x_0 = x_1 = 1$ "die without progeny." It is reasonable—in spite of this inauspicious beginning—to ask whether there exists an x_2—i.e., an integer for which $f(X) \equiv 0 \pmod{p^3}$. Returning temporarily to the original congruence $f(X) = X^2 + X + 7 \equiv 0 \pmod 3$, we see that x is a solution if and only if $x \equiv 1 \pmod 3$. Since $f(1) = 9$, $f(4) = 27$, and $f(7) = 63$, there are precisely three candidates for x_2. The requirement for existence of x_2 is precisely that the linear (in t) congruence $f^{(1)}(x_2)t \equiv -f(x_2)/3^2 \pmod 3$ be solvable. The congruences produced by $x_1 = 1, 7$ aren't solvable (please check), but $x_1 = 4$ produces $9t \equiv -3 \pmod 3$, an identical congruence, which means that $x_1 = 4$ has three children: $x_2 = 4$, 13, and 22, all of which satisfy $f(X) = X^2 + X + 7 \equiv 0 \pmod{3^3}$. Since $f^{(1)}(x_3)t \equiv -f(x_3)/3^3 \pmod 3$ isn't solvable, regardless of the way we choose x_3 (again, please check), we see that the congruence $f(X) = X^2 + X + 7 \equiv 0 \pmod{3^4}$ has no solutions, and so the 3-adic approximation of a solution of $X^2 + X + 7 = 0$ terminates at this point.

PROBLEMS

Solutions (or hints at solutions) to the problems marked with an asterisk (*) are given on page 129.

*1. Referring to page 90, Example 4.2.2, verify the following statements.
   *(a) Neither of the 3-adic second approximations $x_1 = 1, 7$ to a solution of $f(X) = X^2 + X + 7 = 0$ can be refined further to yield a 3-adic third approximation to a solution of this equation.
   *(b) The 3-adic second approximation $x_1 = 4$ to a solution of $f(X) = X^2 + X + 7 = 0$ can be refined to yield a 3-adic third approximation to this equation, and in fact it has precisely three progeny of this type, which can be taken to be $x_2 = 4$, 13, and 22.
   *(c) No 3-adic fourth approximations to the equation $f(X) = X^2 + X + 7 = 0$ exist; i.e., there is no integer x_3 satisfying $f(x_3) \equiv 0 \pmod{3^4}$.

2. Solve $f(X) = X^2 + X + 7 \equiv 0 \pmod{3^3}$, using the technique described near the end of Section 3.3 (see pages 56–57).

*3. Assume that x_0 is a solution of the polynomial congruence $f(X) \equiv 0 \pmod p$. Prove that a complete list of candidates for solutions of $f(X) \equiv 0 \pmod{p^2}$ can be taken to be $\{x_0, x_0 + p, x_0 + 2p, \ldots, x_0 + (p - 1)p\}$.

***4.** Solve the following polynomial congruences, using the theory developed to date in the text:
(a) $X^3 + X^2 - 4 \equiv 0 \pmod{7^3}$.
***(b)** $X^3 - 3X^2 + 27 \equiv 0 \pmod{1{,}125}$.
(c) $X^3 + X^2 - 5 \equiv 0 \pmod{343}$.

5. Verify formula (1) on page 89.

Definition 4.2.1: An integer x_0 is said to be a **nonsingular** solution of the polynomial congruence $f(X) \equiv 0 \pmod{p^e}$, p being a prime, provided that $p \nmid f^{(1)}(x_0)$ and of course that $f(x_0) \equiv 0 \pmod{p^e}$.

---

6. Find all the nonsingular solutions of the quadratic congruence $X^2 \equiv b \pmod{p^e}$.
7. (a) Assume that the polynomial congruence $f(X) \equiv 0 \pmod{p}$ has exactly s distinct solutions, all nonsingular. Prove that the same is true of the polynomial congruence $f(X) \equiv 0 \pmod{p^e}$ for every integer $e \geqslant 2$.
(b) Assume that $d|(p - 1)$, where p is a prime. Prove that the polynomial congruence $X^d - 1 \equiv 0 \pmod{p^e}$ has exactly d solutions for every $e \geqslant 1$.

***8.** Example 4.2.1 on page 89 can be regarded as finding a **7-adic square root** of the integer 11. Find the other 7-adic square root of 11.

***9.** Prove that the binomial coefficient $\binom{i}{k}$ is an integer.

†10. Let $f(X) = X^3 + 396X^2 - 111X + 38$. Prove that the congruence $f(X) \equiv 0 \pmod{3^e}$ has precisely 3^2 solutions for every $e \geqslant 5$.

4.3 Polynomial Congruences with Respect to a General Modulus; i.e., $f(X) \equiv 0 \pmod{m}$

I want to do almost nothing on this topic beyond indicating how we are able to climb the rest of the way up our ladder. As mentioned at the outset of Chapter 4, the problem of solving the polynomial congruence $f(X) \equiv 0 \pmod{m}$ is equivalent to the system of congruences

$$f(X) \equiv 0 \pmod{p_1^{e_1}}$$
$$f(X) \equiv 0 \pmod{p_2^{e_2}}$$

$$\vdots$$

$$f(X) \equiv 0 \pmod{p_k^{e_k}}$$

where $p_1^{e_1} \cdot p_2^{e_2} \cdot \ldots \cdot p_k^{e_k}$ is the canonical decomposition of m. Now suppose that $\{b_{i1}, b_{i2}, \ldots, b_{is}\}$ is a complete set of solutions for $f(X) \equiv 0 \pmod{p_i^{e_i}}$, with $1 \leqslant i \leqslant k$—i.e., $X \equiv b_{i1}$ or b_{i2} or ... or $b_{is} \pmod{p_i^{e_i}}$. The upshot of all of this is that we wind up with $s_1 \cdot s_2 \cdot \ldots \cdot s_k$ systems of linear congruences, each of the type

$$x \equiv b_{1j} \pmod{p_1^{e_1}}$$
$$x \equiv b_{2j} \pmod{p_2^{e_2}}$$
$$\vdots$$
$$x \equiv b_{kj} \pmod{p_k^{e_k}}$$

The Chinese Remainder Theorem tells us that each such system has precisely one solution modulo m, and so all solutions of the original congruence can be obtained.

PROBLEM

A hint at a solution to this problem is given on page 130.

*1. Find all solutions of the polynomial congruence $f(X) = X^3 + 3X^2 + 27 \equiv 0 \pmod{1{,}125}$.

NOTES TO CHAPTER 4

1. The discussion in Section 4.2 owes more to the treatment in LeVeque's *Fundamentals of Number Theory* than to any other single source. See, specifically, pages 66–68 of that text. LeVeque also has a more advanced treatment in his Section 3.5, The p-adic fields (pages 70–77), where he goes into the analysis and topology of the situation.
2. In conjunction with Example 4.2.1 (page 90) and the follow-up problem 8 (page 91), there is Hermann Weyl's *Algebraic Theory of Numbers,* in particular pages 88–90. Weyl's book, however, is at a much more advanced level than the present text.

5 Carl Friedrich Gauss' Law of Quadratic Reciprocity

INTRODUCTION

The object of this final chapter is to initiate a study of one of the most beautiful theorems in all of mathematics, Gauss' Law of Quadratic Reciprocity.

If p is any arbitrary but fixed odd prime, then we can—by brute force, if necessary—find all integers b relatively prime to p for which the quadratic congruence $X^2 \equiv b \pmod{p}$ is solvable. These values of b are the so-called quadratic residues modulo p (see Example 3.7.1 on page 74, and more generally Definition 3.7.1 on page 74). So here the question is: which integers are quadratic residues of an arbitrary but fixed odd prime p? This problem is finite in nature because, for example, we may discover a complete set of values of quadratic residues of p that are pairwise incongruent modulo p by computing $1^2, \ldots, (p-1)^2$ (please verify this statement). Actually, it suffices to go "halfway"—see page 97, problem 1. But suppose that, instead of preassigning p and looking for all b, we consider the inverse problem in which we preassign some integer b and go hunting for all odd primes p of which b is a quadratic residue. This problem is, at least a priori, an infinite problem because there are infinitely many primes. However, maybe it only seems to be infinite; i.e., perhaps there are really only finitely many primes for which a given b can be a quadratic residue. The set of primes we are looking for is precisely the set of primes entering the canonical decompositions of the elements of the infinite set $\{-b, 1^2 - b, 2^2 - b, 3^2 - b, \ldots\}$. Is it obvious that the set of primes in question is infinite? Is it true? Can you prove it? (See page 99, problem 8†(a) and (c).)

The Law of Quadratic Reciprocity enables us to solve completely (in a sense to be made precise later) this infinite inverse problem by relating it to the original finite problem, thus establishing the finitude of the inverse problem! All of this sounds somewhat vague at this point, but the main message is (I hope) already clear: the Law of Quadratic Reciprocity has the effect of transforming a problem that in its original formulation is inherently infinite into a problem that is most definitely finite—surely a remarkable feat!

5.1 Technical Results Needed for This[1] Proof of the Law of Quadratic Reciprocity

We collect in this section all the material needed to make the particular proof of Gauss' Law of Quadratic Reciprocity presented in Section 5.2.[1] The proof will then be quite quickly obtained, but that should not in any way be construed to mean that the theorem itself is easy to come by! We start by introducing the so-called Legendre symbol—an extremely useful notational device, but (as we shall soon see) much more than mere notation.

Definition 5.1.1: If $(b, p) = 1$, where p is any arbitrary but fixed odd prime, then the **Legendre symbol,** denoted by (b/p), is defined as follows:

1. $(b/p) = +1$ if b is a quadratic residue of p.
2. $(b/p) = -1$ if b is a quadratic nonresidue of p.

As always, we shall try to get some feel for the current definition by looking at some numerical examples and then establishing some of the basic properties of the Legendre symbol.

Example 5.1.1: $(1/7) = +1$, $(2/7) = +1$, $(3/7) = -1$, $(4/7) = +1$, $(5/7) = -1$, and $(6/7) = -1$. It is obvious (?) that, if $b \equiv c \pmod{p}$, then $(b/p) = (c/p)$; so there is no loss of generality in assuming that $1 \leqslant b \leqslant (p - 1)$.

Example 5.1.2: $(1/11) = (4/11) = (9/11) = (16/11) = (5/11) = (25/11) = (3/11) = +1$, and $(2/11) = (6/11) = (7/11) = (8/11) = (10/11) = -1$. We can see that the number of quadratic residues modulo 11 is equal to the number of quadratic nonresidues modulo 11 (compare page 74, Example 3.7.1) and that $2 + 6 + 7 + 8 + 10 > 1 + 3 + 4 + 5 + 9$, for instance. (See also page 99, problem 8†(b).)

[1] See page 104, note 1.

List of Some of the Basic Properties of the Legendre Symbol

1. $b \equiv c \pmod p$ implies $(b/p) = (c/p)$ (**periodicity**).
2. $(bc/p) = (b/p)(c/p)$ for all choices of the integers b and c, subject to the constraints $(b, p) = (c, p) = 1$ (**multiplicativity**).
3. $(b/p) \equiv b^{(p-1)/2} \pmod p$ for all odd primes p and all integers b such that $p \nmid b$ (**Euler's Criterion**). (Compare with page 75, Theorem 3.7.1—Euler's Criterion. See also page 98, problem 2, particularly the hint.)

Proof of Property 3: Fermat's Little Theorem (see page 50) tells us that $b^{p-1} \equiv 1 \pmod p$; i.e., $(b^{(p-1)/2})^2 \equiv 1^2 \pmod p$. The divisibility way of saying this is $p|(b^{(p-1)/2} - 1)(b^{(p-1)/2} + 1)$, from which we conclude that $p|(b^{(p-1)/2} - 1)$ or $p|(b^{(p-1)/2} + 1)$. Converting back to the congruence notation, we see that $b^{(p-1)/2} \equiv \pm 1 \pmod p$. However, specializing Euler's Criterion (see page 75)—which provides a criterion for what it takes to be an nth power residue modulo m—we find that b is a second-power residue (i.e., a quadratic residue) modulo p if and only if $b^{(p-1)/2} \equiv +1 \pmod p$. The component parts of the proof are now all assembled.

Remark: The Legendre symbol can be thought of very elegantly with the help of a little abstract algebra. For instance, $f(b) = (b/p)$ can be thought of as defining a map from F_p^* (the multiplicative group of nonzero elements of the Galois field of p elements) onto the multiplicative group $\{\pm 1; \cdot\}$. The multiplicative property of the Legendre symbol, for instance, becomes the statement that this map f is a group homomorphism.

We next collect Gauss' Lemma.

Gauss' Lemma: Assume that p is an odd prime and that $(b, p) = 1$. Let $S = \{1 \cdot b, 2 \cdot b, \ldots, ((p - 1)/2) \cdot b\}$. If n denotes the number of elements of S that have least positive[1] residue modulo p in excess of $p/2$, then $(b/p) = (-1)^n$.

Before making the proof, it may be worthwhile to give a numerical example to make sure that we understand what Gauss' Lemma is telling us.

Example 5.1.3: Let's take $p = 17$ and $b = 5$. The set S becomes $\{5, 10, 15, 20, 25, 30, 35, 40\}$, and the induced set of least positive residues modulo 17[1] is $\{5, 10, 15, 3, 8, 13, 1, 6\}$; so $n = 3$, and hence 5 is a quadratic nonresidue modulo 17.

Proof—"upon Reflection": Let $\{R_1, R_2, \ldots, R_n; r_1, r_2, \ldots, r_{((p-1)/2)-n}\}$ denote the set of least positive residues modulo p[1] emanating from the original

[1]This is simply what you get from the Division Algorithm.

set S. The notation encodes the assumption that there are precisely n of these least positive residues—namely, $R_1, R_2, \ldots, R_n$—satisfying $R_i > p/2$, $1 \leq i \leq n$. We now form the new set $\{p - R_1, p - R_2, \ldots, p - R_n; r_1, r_2, \ldots, r_{((p-1)/2)-n}\}$ by "reflecting" each R_i about the point $p/2$; I claim that this new set is the set $\{1, 2, 3, \ldots, (p - 1)/2\}$ in some order (please verify—see page 98, problem 4). Having done the "geometrical" part of the proof, we now do the algebraic part: $(p - R_1) \cdot (p - R_2) \cdot \ldots \cdot (p - R_n) \cdot r_1 r_2 \cdot \ldots \cdot r_{((p-1)/2)-n} = 1 \cdot 2 \cdot \ldots \cdot (p - 1)/2$ implies $(-1)^n R_1 R_2 \cdot \ldots \cdot R_n r_1 r_2 \cdot \ldots \cdot r_{((p-1)/2)-n} \equiv 1 \cdot 2 \cdot 3 \cdot \ldots \cdot (p - 1)/2 \pmod p \equiv 1 \cdot 2 \cdot 3 \cdot \ldots \cdot ((p - 1)/2) \cdot b^{(p-1)/2} \pmod p$, so $(-1)^n(1 \cdot b)(2 \cdot b) \cdot \ldots \cdot ((p - 1)/2) \cdot b) \equiv 1 \cdot 2 \cdot \ldots \cdot (p - 1)/2 \pmod p$. Hence, $(-1)^n \cdot b^{(p-1)/2} \equiv 1 \pmod p$, and thus finally $b^{(p-1)/2} \equiv (-1)^n \pmod p$. However, property 3 of the Legendre symbol (see page 95) tells us that $b^{(p-1)/2} \equiv (b/p) \pmod p$, and thus $(b/p) \equiv (-1)^n \pmod p$, and we have $(b/p) = (-1)^n$ as desired.

We need one more technical lemma to pave the way for the particular proof of the Law of Quadratic Reciprocity that we have in mind. This result can be made to depend heavily on Gauss' Lemma, as we shall see.

Technical Lemma: Assume that p is an odd prime and that $(b, 2p) = 1$. Then

$$(b/p) = (-1)^t, \text{ where } t = \sum_{k=1}^{\frac{p-1}{2}} [kb/p].$$[1]

Proof: The Division Algorithm, applied to the pair of integers kb and p, gives $kb = p[kb/p] +$ (the least positive remainder of kb modulo p). Summing over the relevant values of k produces the equation

$$\sum_{k=1}^{\frac{p-1}{2}} kb = \sum_{k=1}^{\frac{p-1}{2}} p[kb/p] + \sum_{i=1}^{n} R_i + \sum_{j=1}^{\frac{p-1}{2}-n} r_j \tag{1}$$

where the symbols R_i and r_j have the same meanings as on page 95. From the proof of Gauss' Lemma we recapture

$$\sum_{k=1}^{\frac{p-1}{2}} k = \sum_{i=1}^{n} (p - R_i) + \sum_{j=1}^{\frac{p-1}{2}-n} r_j \,. \tag{2}$$

[1]The notation [ ] denotes the greatest integer function; see page 7, problem 14.

Subtracting equation (2) from equation (1) produces equation (3):

$$\sum_{k=1}^{\frac{p-1}{2}} (b-1)k = (b-1)\sum_{k=1}^{\frac{p-1}{2}} k$$

$$= (b-1)(p^2-1)/8 \tag{3}$$

$$= p\{\sum_{k=1}^{\frac{p-1}{2}} [kb/p] - n\} + 2\sum_{i=1}^{n} R_i$$

(please verify!).

Hence,

$$(b-1)((p^2-1)/8) \equiv \sum_{k=1}^{\frac{p-1}{2}} [kb/p] - n \pmod 2. \tag{4}$$

Since b is odd by assumption, we get

$$\sum_{k=1}^{\frac{p-1}{2}} [kb/p] \equiv n \pmod 2$$

and the result follows upon our applying Gauss' Lemma.

Corollary 5.1.1: $(2/p) = (-1)^{(p^2-1)/8}$.

Proof: If $b = 2$, everything proceeds as in the proof of the Technical Lemma, except for the last step. Substituting $b = 2$ in equation (4) gives

$$(p^2 - 1)/8 \equiv \sum_{k=1}^{\frac{p-1}{2}} [2k/p] - n \pmod 2$$

which is better written $n \equiv (p^2 - 1)/8 \pmod 2$ (justification?). The result now follows from Gauss' Lemma.

PROBLEMS

Solutions (or hints at solutions) to the problems marked with an asterisk are given on page 130. Assume throughout this problem set that p is an odd prime.

***1.*(a)** Prove that the numbers $1^2, 2^2, \ldots, ((p-1)/2)^2$ are pairwise incongruent modulo p.

*(b) Prove that, if b is a quadratic residue modulo p, then $b \equiv i^2 \pmod{p}$ for some integer i satisfying $1 \leqslant i \leqslant (p - 1)/2$.
(c) Conclude (from parts (a) and (b) of this problem) that there are precisely $(p - 1)/2$ quadratic residues modulo p and that they are pairwise incongruent modulo p.
*(d) Find a complete set of quadratic residues in case $p = 13$, using what you have learned from parts (a), (b), and (c) of this problem.
*(e) Find a complete set of quadratic nonresidues with respect to the modulus 11.

*2. (a) Prove the periodic property of the Legendre symbol—i.e., $(b/p) = (c/p)$ whenever $b \equiv c \pmod{p}$.
*(b) Does $(b/p) = (c/p)$ force $b \equiv c \pmod{p}$?
(c) Prove the multiplicative property of the Legendre symbol—i.e., $(bc/p) = (b/p)(c/p)$ for all integers b and c for which the symbols (b/p) and (c/p) are defined.

[HINT (strong!): For parts (a) and (c), you can make good use of Euler's Criterion (property 3 of the Legendre symbol on page 95).]

*3. Verify Gauss' Lemma in the following situations:
*(a) $p = 13$, $b = 10$.
(b) $p = 23$, $b = 15$.

4. Referring to the proof of Gauss' Lemma (page 95), establish the following set equality:

$$\{p - R_1, p - R_2, \ldots, p - R_n; r_1, r_2, \ldots, r_{((p-1)/2)-n}\} = \{1, 2, 3, \ldots, (p - 1)/2\}.$$

*5. Referring to the proof of the Technical Lemma, page 96,
(a) Verify equation (3) on page 97.
*(b) Prove that $(p^2 - 1)/8 \equiv \sum_{k=1}^{\frac{p-1}{2}} [2k/p] - n \pmod{2}$ is equivalent to the congruence $n \equiv (p^2 - 1)/8 \pmod{2}$.

***6. The quadratic character of (-1).**
*(a) Use property 3 on page 95, of the Legendre symbol to show that $(-1/p) = (-1)^{(p-1)/2}$. Note that this is an actual equality, not just a congruence modulo p.
*(b) Prove that the quadratic congruence $X^2 \equiv -1 \pmod{p}$ is solvable if and only if $p = 2$ or $p \equiv 1 \pmod{4}$. (Compare page 58, problem 9.)
*(c) Are there infinitely many primes for which (-1) is a quadratic residue? Justify your answer.

***7. The quadratic character of 2.**
(a) Prove that the quadratic congruence $X^2 \equiv 2 \pmod{p}$ is solvable if and only if the prime p is equal to 2 (trivial) or $p \equiv \pm 1 \pmod{8}$.

*(b) Are there infinitely many primes p for which 2 is a quadratic residue modulo p? Prove that your answer is correct.

[HINTS: (1) Apply Corollary 5.1.1 on page 97 to solve part (a). (2) Mimic Euclid in order to solve part (b); see, for instance, page 24, problem 12(a) and (b).]

8.†(a) Consider the set $\{-b, 1^2 - b, 2^2 - b, 3^2 - b, \ldots\} = B$. Is it true that the set of primes entering the canonical decompositions of the elements of the set B is infinite, regardless of the choice of the integer b?

*(b) Is it always correct to claim that $\Sigma n > \Sigma r$, where the first sum is understood to extend over all the quadratic nonresidues of the odd prime p, while the second sum is understood to extend over all the quadratic residues of that same odd prime p?

(c) Prove that your answers to problem 8(a) and (b) are correct.

5.2 Gauss' Law of Quadratic Reciprocity

Now that we have assembled, in Section 5.1, some of the basic properties of the Legendre symbol, together with Gauss' Lemma and the closely associated Technical Lemma, we are in a position to prove the deepest result in this text—namely, Gauss' Law of Quadratic Reciprocity.

Theorem 5.2.1 – Gauss' Law of Quadratic Reciprocity: If p and q are distinct odd primes, then

$$(p/q)(q/p) = (-1)^{((p-1)/2)((q-1)/2)}. \tag{1}$$

Remark on Notation: The factors on the left-hand side of this equation are both Legendre symbols, but $(p - 1)/2$ and $(q - 1)/2$ are not. I have deliberately been very careful not to write $(p - 1/2)$, so as to try to avoid any possible confusion. Another way to remember not to interpret $((p - 1)/2)$—even if it were written $(p - 1/2)$—as a Legendre symbol is by looking back at page 94, Definition 5.1.1: the Legendre symbol (b/p) was only defined for p an odd prime.

Proof: Let $S = \{(x, y) \in Z \times Z | 1 \leqslant x \leqslant (p - 1)/2,\ 1 \leqslant y \leqslant (q - 1)/2\}$; i.e., S is the set of all **lattice points** (a point (x, y) with both coordinates integral) in the rectangle $[1, (p - 1)/2] \times [1, (q - 1)/2]$ and thus has $((p - 1)/2)$ $((q - 1)/2)$ elements. We partition the set S into two sets S_1 and S_2 by defining

$S_1 = \{(x, y) \in Z \times Z | 1 \leqslant x \leqslant (p - 1)/2, 1 \leqslant y \leqslant qx/p\}$ and $S_2 = S - S_1$. If x is preassigned—subject to the constraint $1 \leqslant x \leqslant (p - 1)/2$—then the number of available y values is $[qx/p]$. Hence, the total number of elements in the set S_1 is given by the formula $\sum_{x=1}^{\frac{p-1}{2}} [qx/p]$. In like manner, the number of elements of the set S_2 is given by the formula $\sum_{y=1}^{\frac{q-1}{2}} [py/q]$ (please verify!), with the result that $\sum_{x=1}^{\frac{p-1}{2}} [qx/p] + \sum_{y=1}^{\frac{q-1}{2}} [py/q] = ((p - 1)/2)((q - 1)/2)$. The Technical Lemma (page 96) now immediately gives $(p/q)(q/p) = (-1)^{((p-1)/2)((q-1)/2)}$, and we are done!

Remark: The total exponent on (-1) in Gauss' Law of Quadratic Reciprocity is $((p - 1)/2)((q - 1)/2)$, which is an even integer whenever at least one of the conditions $p \equiv 1 \pmod 4$, $q \equiv 1 \pmod 4$ is satisfied. Hence, in all such cases, we are able to conclude that $(p/q) = (q/p)$. In the case remaining—i.e., $p \equiv q \equiv 3 \pmod 4$—we have $(p/q) = -(q/p)$. Thus, in all (four) cases, we have a simple rule to follow in order to "invert" Legendre symbols in which the "numerator" is an odd prime. This fact, in conjunction with the basic properties of the Legendre symbol, makes it very easy (and fun!) to evaluate Legendre symbols.

Example 5.2.1: Evaluate $(-1{,}890/101)$; i.e., determine whether $-1{,}890$ is a quadratic residue modulo 101—or, in other words, determine whether the quadratic congruence $X^2 \equiv -1{,}890 \pmod{101}$ is solvable. Various methods for solving such a problem exist—some of them more efficient than others. Here is one way to get the job done (I'm not claiming that this solution is the most efficient!). $1{,}890 = 2 \cdot 3^3 \cdot 5 \cdot 7$, so $(-1{,}890/101) = (-1/101) \cdot (2/101) \cdot (3/101)^3 \cdot (5/101) \cdot (7/101)$, by the multiplicativity of the Legendre symbol. Since $(b/p) = \pm 1$, regardless of the odd prime p and the choice of the integer b relatively prime to p, $(3/101)^3 = (3/101)$. The special formula $(-1/p) = (-1)^{(p-1)/2}$ (see page 98, problem 6(a)) gives $(-1/101) = (-1)^{(101-1)/2} = +1$, while the other special formula $(2/p) = (-1)^{(p^2-1)/8}$ (see page 97, Corollary 5.1.1) gives $(2/101) = (-1)^{(101^2-1)/8} = -1$; so now we have $(-1{,}890/101) = -(3/101) \cdot (5/101) \cdot (7/101)$. Next, $(3/101) = (101/3)$ (see the remark on page 100) $= (2/3) = -1$, where we have used the periodicity property of the Legendre symbol (see page 95, property 1). Likewise, $(5/101) = (101/5) = (1/5) = +1$ and $(7/101) = (101/7) = (3/7) = -1$, and our final answer is $(-1{,}890/101) = -1$; i.e., the congruence $X^2 \equiv -1{,}890 \pmod{101}$ is not solvable.

Our next example is much more important than Example 5.2.1 from a theoretical point of view. It makes it clear that the inverse problem mentioned in the introduction to Chapter 5 is, in general, infinite. Even more important, it indicates that the answer to the inverse problem has a strong internal structure. These rather vague statements will be made precise in the remarks following Example 5.2.2.

Example 5.2.2: Characterize those odd primes p of which 7 is a quadratic residue. If p is any odd prime other than 7, the Law of Quadratic Reciprocity gives $(7/p) = (p/7) \cdot (-1)^{((7-1)/2)((p-1)/2)}$, so that $(7/p) = +1$ if and only if $(p/7) = (-1)^{(p-1)/2}$. If $p \equiv 1 \pmod 4$, then $(p/7) = +1$, and p can be congruent to 1, 2, or 4 (mod 7). Hence all odd primes $p \equiv 1, 9$, or 25 (mod 28) have 7 as a quadratic residue. If $p \equiv 3 \pmod 4$, then $(p/7) = -1$, and p can be congruent to 3, 5, or 6 modulo 7. Hence, the rest of the odd primes p having 7 as a quadratic residue are exactly those odd primes $p \equiv 3, 19$, or 27 (mod 28).

Remarks on the extreme significance of Example 5.2.2: Dirichlet's famous Theorem on Primes in Arithmetic Progression (page 34, note 4) tells us, for instance, that there are infinitely many primes satisfying $p \equiv 19 \pmod{28}$, because the set of all nonnegative integers z satisfying $z \equiv 19 \pmod{28}$ is an arithmetic progression in which the common difference (28) is relatively prime to the first element (19). Hence, the previously mentioned inverse problem is truly infinite. More precisely, we have $(7/p) = +1$ if and only if $p \equiv 1, 3, 9, 19, 25$, or 27 (mod 28); so that, if S is the set of all odd primes having 7 as a quadratic residue, then S consists precisely of all the odd primes in the six arithmetic progressions just mentioned. Thus the solution set S of our original problem has a very strong internal arithmetic structure. Finally, notice that $6 = \phi(28)/2$ and that, ab initio, the prime p had twelve (because $\phi(28) = 12$) arithmetic progressions to choose from. Hence—in this sense—$(7/p) = +1$ occurs for one-half of all primes p. This rather vague statement can be made rigorous by using more advanced analytic results. Short of that, we content ourselves with one last result, which is a very satisfactory and general extension of Example 5.2.2.

Theorem 5.2.2: Assume that p and q are distinct odd primes, with $q > p$.

1. If $p \equiv 1 \pmod 4$ and $q = pQ + r$, with $0 < r < p$, then $(p/q) = (r/p)$.
2. If $p \equiv 3 \pmod 4$, then again $(p/q) = (r/p)$, where (in this case) r is uniquely determined via the following constraints:

$$q = (4p)Q + r,\ 0 < r < 4p, \text{ and } r \equiv 1 \pmod 4.$$

Proof: Part 1 is immediate: $(p/q) = (q/p)$ by the Law of Quadratic Reciprocity because $p \equiv 1 \pmod 4$, and $(q/p) = (r/p)$ by the periodicity of the Legendre symbol.

Our first job in establishing part 2 is to prove the existence and uniqueness of the claimed representation of q in terms of p. If we write $q = (4p)M + R$, where $0 < R < 4p$ via the Division Algorithm, then $r = R$ if $R \equiv 1 \pmod 4$. Otherwise, $r = 4p - R$. If $q = (4p)Q + r$, then $(q/p) = (r/p)$ (periodicity) and $(q/p) = (p/q)$ (reciprocity in conjunction with $q \equiv r \equiv 1 \pmod 4$). If $q = (4p)Q - r$, then $(q/p) = (-r/p) = (-1/p)(r/p) = -(r/p)$ (recall $p \equiv 3 \pmod 4$), and $(q/p) = -(p/q)$ because $q \equiv 3 \pmod 4$ also.

Example 5.2.3: To illustrate the utility of Theorem 5.2.2, let's take $p = 19$, so that $p \equiv 3 \pmod 4$. The set of all integers r satisfying $0 < r < 4p = 76$, $r \equiv 1 \pmod 4$, and $(r/19) = +1$ is $\{1, 5, 9, 17, 25, 45, 49, 61, 73\}$. Theorem 5.2.2 allows us to conclude that the odd primes $q > p = 19$ having 19 as a quadratic residue are precisely all primes $q > 19$ of the form $q = (4p)Q + r = 76Q + r$, where r runs through all elements of the set just determined. Here we have eighteen arithmetic progressions and $18 = \phi(76)/2 = \phi(4 \cdot 19)/2$. Please compare with the remarks on the extreme significance of Example 5.2.2 (page 101).

PROBLEMS

Solutions (or hints at solutions) to the problems marked with an asterisk are given on page 130.

*1. (a) How many solutions does the congruence $X^2 \equiv 37 \pmod{2{,}772}$ have?

*(b) Evaluate (501/773) and (503/773).

*2. Prove that, if the distinct primes p and q satisfy $p \equiv q \equiv 3 \pmod 4$ and if $X^2 \equiv p \pmod q$ has no solutions, then $X^2 \equiv q \pmod p$ has two solutions.

*3. Assume that b is a quadratic nonresidue of each of the odd primes p and q. Is the congruence $X^2 \equiv b \pmod{pq}$ solvable? Justify your answer.

4.(a) Characterize the primes p satisfying $(5/p) = +1$.

(b) Characterize the primes p satisfying $(10/p) = -1$.

5. (a) Prove that there are infinitely many primes congruent to $+1$ modulo 3.

(b) Prove that there are infinitely many primes congruent to -1 modulo 3. [HINT: Investigate those primes for which (-3) is a quadratic residue.]

***6.** Assume that p is an odd prime and that $p \nmid b$. Prove that $\sum_{k=1}^{p-1} (kb/p) = 0$.

***7.** Assume that $p = 4^k + 1$ for some positive integer k. Prove that p is a prime if and only if $3^{(p-1)/2} \equiv -1 \pmod p$. (Compare page 74, problem 5.)

8. Prove that there are no integers x and y for which $y^2 = x^3 + 7$. [HINT: If (x, y) were a solution, x would have to be odd (why?), so write $x = 2z + 1$. Now factor $y^2 + 1$ as a polynomial in z.]

9. Characterize the prime powers p^e for which there exist integers x and y satisfying $p \nmid xy$ and $x^2 + y^2 \equiv 0 \pmod{p^e}$.

***10.** Prove that the product of the quadratic residues of the odd prime p is congruent to -1 or $+1$ modulo p, according as $p \equiv +1 \pmod 4$ or $p \equiv -1 \pmod 4$. [HINT: Let g be a primitive root of p, and express the quadratic residues of p in terms of g.]

***11.** Prove each of the following:

***(a)** If p is an odd prime and if $q = 4p + 1$ is also a prime, then 2 is a primitive root of the prime q.

(b) If $p \equiv 1 \pmod 4$ is a prime and if $q = 2p + 1$ is also a prime, then 2 is a primitive root of the prime q.

(c) If $p \equiv -1 \pmod 4$ is a prime and if $q = 2p + 1$ is also a prime, then -2 is a primitive root of the prime q.

(d) If p and $q = 2p + 1$ are both odd primes, then -4 is a primitive root of q.

***12. The Geometry of the Law of Quadratic Reciprocity.** In conjunction with the proof of the Law of Quadratic Reciprocity (page 99),

***(a)** Justify the statement, "If x is preassigned—subject to the constraint $1 \leqslant x \leqslant (p - 1)/2$—then the number of available y values is $[qx/p]$."

(b) Show that $^{\#}(S_2) = \sum_{y=1}^{\frac{q-1}{2}} [py/q]$.

***(c)** Make a sketch of the plane analytic geometry attendant upon this proof. Are there any lattice points on the line $y = (p/q)x$? Why?

***13.†** Assume that p and q are distinct odd primes and that the **exponential Diophantine equation** $1 + q + q^2 + \ldots + q^{x-1} = p^y$ has a solution in positive integers with $x > 3$. It is known that $x = \text{ord}_p q$ and that $y = u \cdot \text{ord}_q p$ for some positive integer u. Must it be the case that $u = 1$; i.e., is there some form of reciprocity here?

***14.†** Prove that $1 + x + x^2 + x^3 + x^4 = y^2$ has $x = 3$, $y = 11$ as its only integral solution.

NOTES TO CHAPTER 5

1. I emphasized—in the title of Section 5.1—the fact that I was rounding up the technical results needed for the particular proof of the Law of Quadratic Reciprocity to be presented. The history of this famous theorem makes for fascinating reading. Gauss—at the age of seventeen!—was the first to give a complete proof, although Legendre had stated the result previously. Gauss eventually gave eight different proofs, and Murray Gerstenhaber has an article in *The American Mathematical Monthly,* Volume 70 (1963), 397–98 entitled "The 152-nd Proof of the Law of Quadratic Reciprocity." For more detail see LeVeque's *Fundamentals of Number Theory*, pages 107–8.
2. See Borevich and Shafarevich's *Number Theory,* Theorem 4, pages 346–47, for information relating to problem 8(b) and (c) on page 99.
3. The particular proof of the Law of Quadratic Reciprocity given on page 99 can be looked at in a highly geometric way; see, for instance, pages 76–77 of Hardy and Wright's classic *An Introduction to the Theory of Numbers,* Fifth Edition.
4. Example 5.2.2 (page 101) through the end of Example 5.2.3 (page 102), are precursors of, in Harvey Cohn's words, "an esoteric synthesis of number theory, analysis, and algebra called 'class-field theory.'" Cohn's text, *A Second Course in Number Theory,* is a good reference for more advanced study along these lines.

I Appendix I—Basics

My purpose here is to collect mathematical material designed to make this text maximally self-contained. If the reader encounters difficulty in the body of the text, this appendix is designed to help remove that difficulty. The material here is arranged in the order in which it is encountered in the text proper.

1. EQUIVALENCE RELATIONS

If S is any set, then the **Cartesian product** of S with itself, denoted $S \times S$, consists of all ordered pairs of elements of S—i.e., all objects of the form (s, s'), where s and s' are completely arbitrary elements of S. A **binary relation** on S, call it "twiddle" and denote it by $\sim$, is any subset of $S \times S$; we say that (s, s') is an element of $\sim$ if and only if $s \sim s'$. In order to try immediately to avoid confusion, consider the following example.

Example I.1.1: Divisibility (see page 1, Definition 1.1.1) is a binary relation on the set Z. The element (b, c) of $Z \times Z$ belongs to the binary relation $|$ if and only if $b|c$.

Definition I.1.1: The binary relation $\sim$ on the set S is said to be an **equivalence relation** on S if and only if it satisfies the following three properties:

1. $s \sim s$ for all s in S (the **reflexive** property). [NOTE: $s \sim s$ means that (s, s) is an element of the binary relation $\sim$.]
2. If $s \sim s'$, then $s' \sim s$, for each $s, s' \in S$ (the **symmetric** property).
3. If $s \sim s'$ and $s' \sim s''$, then $s \sim s''$, for each $s, s', s'' \in S$ (the **transitive** property).

Example I.1.1 Revisited: Divisibility is transitive but isn't symmetric and isn't (quite) reflexive. (See Theorem 1.1.1, part 2, on page 2, and problem 10(a) and (b) on page 6.)

Example I.1.2: Congruence modulo m (which is a binary relation on Z) is an equivalence relation on Z (see Theorem 3.1.1, page 38).

The fact that congruence modulo m is an equivalence relation on Z, whereas divisibility is not, is the single most important reason for the superiority of the congruence notation—one of many testaments to the genius of Gauss.

I want to mention, but not go into, the primary reason for the mathematical success of equivalence relations: if $\sim$ is an equivalence relation on a set S, then $\sim$ **partitions** S; i.e., it chops (set-theoretically) S up into a bunch of nonempty pairwise disjoint subsets whose union is S. An excellent reference for all of this is Herstein's *Topics in Algebra,* Second Edition—specifically Theorem 1.1.1 on page 8.

2. THE PRINCIPLE OF THE LEAST ELEMENT[1]

Strictly speaking, this is an axiom (hypothesis, postulate, assumption) for the set $\{1, 2, 3, \ldots\}$ of positive integers. There are a number of different-sounding, but mathematically equivalent, ways of expressing this principle. We will usually use it in the following form:

Principle of the Least Element: Every nonempty set of positive integers has a least element.

It can be shown that this principle is equivalent to the following one.

Principle of Mathematical Induction: Any set of positive integers that contains the integer 1 and that contains $k + 1$ whenever it contains the positive integer k contains all positive integers.

Exercise I.2.1: The term *least* in the statement of the Principle of the Least Element means least relative to the usual ordering ($<$). The principle is an existence statement. Show that it is also a uniqueness statement—i.e., that if S

[1] Also known as the Well-Ordering Principle.

is our nonempty set of positive integers, then S has one and only one least element. (See page 8, remark 1.)

Example I.2.1: Let's prove, using (mathematical) induction, that the product of any three consecutive positive integers is divisible by 6.

Solution: Let S denote the set of positive integers n for which $6|n(n + 1)(n + 2)$. Our job is to prove that S is the set of all positive integers. Since $6|1(1 + 1)(1 + 2) = 6$, 1 is in our set. According to the Principle of Mathematical Induction, it is now sufficient to show that, if $6|k(k + 1)(k + 2)$ then $6|(k + 1)(k + 2)(k + 3)$. Now $(k + 1)(k + 2)(k + 3) = k(k + 1)(k + 2) + 3(k + 1)(k + 2)$. We are allowed to assume that $6|k(k + 1)(k + 2)$—this is sometimes referred to as the inductive hypothesis—and 6 does divide $3(k + 1)(k + 2)$ (why?), so 6 divides their sum, and our proof by induction is complete!

3. THE IDEA OF A COUNTEREXAMPLE

(See page 4, problem 1.) Suppose that somebody makes the following claim: whenever we have three integers b, c, and d satisfying $b \nmid c$ and $c \nmid d$, then we necessarily have $b \nmid d$. To disprove this claim, it suffices to exhibit (at least) one triple of integers b, c, and d for which $b \nmid c$ and $c \nmid d$, and yet $b \nmid d$ does not hold—i.e., $b|d$. For instance, we could take $b = 2$, $c = 5$, and $d = 8$ in this case. This single triple of integers proves that the original claim is false—it provides a numerical counterexample to the claim.

4. PROOF BY CONTRADICTION

It is virtually impossible to overestimate the usefulness of this particular proof technique in our subject. For some reason or other, proof by contradiction seems to be almost ideally suited to elementary number theory. Early examples in the text include: the fact that $0 \nmid c$, regardless of the choice of the integer c (see page 1); the uniqueness part of the Division Algorithm (see page 3); and the proof of the irrationality of e (see page 6, problem 8). The technique—stated somewhat informally—is as follows: you start by denying the desired conclusion and then, using the rules of deductive logic, argue until you obtain something you know to be false. The fact that you have been able to deduce logically something known to be false means that your original assumption (the denial of the desired conclusion) must have been false. Hence the desired conclusion must be true! In the specific case of the irrationality of e, you start by assuming that e is rational and, for instance, are able to conclude logically that there is an integer strictly between 0 and 1 (something you know to be false).

5. SEQUENCES AND SERIES

The suggested proof of the irrationality of e (page 6, problem 8, again) involves (infinite) series and, hence, a fortiori, (infinite) sequences. An **infinite sequence** $\{s_n\}_{n=1}^{+\infty} = \{f(n)\}_{n=1}^{+\infty}$ is a function f having as its domain of definition the set of positive integers. We say that the **infinite series** $\sum_{n=1}^{+\infty} u_n$ **converges** to the real number S, and we write $\sum_{n=1}^{+\infty} u_n = S$ iff the sequence of partial sums $\{s_n\}_{n=1}^{+\infty}$ converges to S. To say that this sequence converges to S is to say that $\lim_{n \to +\infty} f(n) = S$. Finally, the generic partial sum s_n is defined to be $s_n = \sum_{k=1}^{n} u_k$. The **Maclaurin expansion** of the exponential function is $e^x = \sum_{i=0}^{+\infty} x^i/i!$.

6. BINOMIAL COEFFICIENTS AND THE BINOMIAL THEOREM

If the integers n and k satisfy the continued inequality $0 \leq k \leq n$, then we may define the <u>binomial coefficient</u> $\binom{n}{k}$ to be equal to $\frac{n!}{k!(n-k)!}$, where $r!$—read "r factorial"—is shorthand for the product $1 \cdot 2 \cdot \ldots \cdot r$ of the first r consecutive positive integers. It follows that we can also use the formula $\binom{n}{k} = n(n-1)(n-2) \cdot \ldots \cdot (n-k+1)/(1 \cdot 2 \cdot \ldots \cdot k)$ in order to compute the numerical value of this binomial coefficient (why?). The Binomial Theorem says that $(b+c)^n = \sum_{k=0}^{n} \binom{n}{k} b^{n-k}c^k$ for every positive integer n. Here, we make the agreement to interpret $\binom{n}{0}$ as having value 1; there are various cogent reasons for doing this.

PROBLEMS

Solutions (or hints at solutions) for those problems marked with an asterisk are given on page 131.

***1.** Each of the following statements is true for all positive integers k. In each case, provide a proof by making an application of the Principle of Mathematical Induction.

(a) $2|k(k+1)$ (see page 6, problem 6(a)).
***(b)** $4!|k(k+1)(k+2)(k+3)$.
(c) $8|\{(2k+1)^2 - 1\}$ (see page 6, problem 6(c)).
(d) $(F_k, F_{k+1}) = 1$ (see page 6, problem 9, and page 14, problem 16(a)).
(e) $G_k|G_{k+1}$, where $G_k = 2^{2^k} - 1$ (see page 22, problem 4(b)).
(f) G_k has at least k distinct prime factors (see page 22, problem 4(c)).
***(g)** The product of k consecutive integers is divisible by $k!$ (see page 24, problem 14).
***(h)** $(1 \cdot 1!) + (2 \cdot 2!) + \ldots + (k \cdot k!) = (k+1)! - 1$ (see page 33, problem 5 (c)).

2. Prove the Binomial Theorem (page 108) by induction.

*3. In each of the following, decide whether the given binary relation $\sim$ on the given set S is reflexive, symmetric, transitive:
   (a) S is a completely arbitrary set; $\sim$ is defined via $b \sim c$ if and only if $b = c$.
   ***(b)** S is the set of real numbers. Define $b \sim c$ to mean $bc \neq 0$.
   ***(c)** S is the set of all complex numbers. Define $b \sim c$ to mean $bc = 0$.

*4. ***(a)** Evaluate $\binom{10}{1}$, $\binom{10}{5}$, and $\binom{14}{2}$.

   ***(b)** Is it correct to claim that $\binom{n}{k} = \binom{n}{n-k}$ for all integers n and k satisfying $0 \leqslant k \leqslant n$? Proof?

   ***(c)** Is it correct to claim that the sum of all the binomial coefficients of the form $\binom{n}{k}$, $0 \leqslant k \leqslant n$, is 2^n? Proof?

   (d) Is $\binom{2n}{n}$ the largest among $\binom{2n}{k}$ as k runs through the integers $0 \leqslant k \leqslant 2n$? Proof?

   ***(e)** Use a binomial coefficient to symbolize the total number of bridge hands available.

*5. Analyze the following "proof" of the claim that every horse has 12 legs: there are 2 legs on the left side, 2 on the right side, 2 at the front, 2 at the back, and 1 at each corner.

II Appendix II—Abstract Algebra

My goals here are as follows:

1. Introduce a few of the most basic concepts of abstract algebra.
2. Make a strong case for the claim that elementary number theory and elementary abstract algebra are "isomorphic."
3. Exhibit the extreme utility of the algebraic approach to elementary number theory, through examples.

The most important algebraic concept for us is that of a group.

Definition II.1: A **group** is an ordered pair $\{G; \cdot\}$ consisting of a set of elements G together with a function $\cdot$ from $G \times G$ into G, subject to the following three axioms:

1. $b \cdot (c \cdot d) = (b \cdot c) \cdot d$ for all choices of the elements b, c, and d from the set G (**associativity**).
2. There exists one and only one element of the set G, call it e, such that $e \cdot g = g \cdot e = g$ for every element g of G (**identity**).
3. For each choice of g in G, there exists one and only one element of G, call it g^{-1}, such that $g \cdot g^{-1} = g^{-1} \cdot g = e$ (**inverse**).

I want to try to bring this rather austere-looking definition to life immediately by relating it to material in the text.

Example II.1: Let $m > 1$ be our modulus. Take for G any complete residue system modulo m, and take the function to be addition modulo m. Then the pair

$\{G; \cdot\}$ constitutes a group. A simple specific example would be $m = 6$, $G = \{0, 1, 2, 3, 4, 5\}$ and $\cdot =$ addition modulo 6; so that, for instance, $3 \cdot 5 = 2$ because $3 + 5 \equiv 2 \pmod 6$. Rather than give a formal proof (see page 117, problem 1(a)), I will simply ask you to notice that associativity is "inherited" from the associativity of (ordinary, garden variety) addition of integers and that the role of e is played by 0.

Example II.2: Again, let $m > 1$ be a given modulus. This time, take G to be any reduced residue system modulo m, say R, and take the function to be multiplication modulo m. If $m = 10$ and $R = \{-3, -1, +1, +3\}$, then, for instance, $(-3) \cdot (-3) = -1$ because $(-3)(-3) \equiv -1 \pmod{10}$. Since ordinary multiplication is associative, so also is multiplication modulo m. The role of e is played by the element of R that is congruent to 1 modulo m. Every r in R has its own inverse (uniquely determined) because the linear congruence $rX \equiv 1 \pmod m$ is solvable. (See page 117, problem 1(b).)

When we are thinking of a group in elementary number theory, it is almost invariably a group of the type illustrated in Examples II.1 and II.2.

Remarks on Definition II.1:

1. The function $\cdot$ of Definition II.1 is often referred to as a **binary operation** on the set G: for each ordered pair (g, g') of elements of G, the function $\cdot$ produces for us the object $g \cdot g'$, which is required under the definition to be an element of G. Sometimes we say that the set G is **closed** with respect to the binary operation $\cdot$.
2. The function $\cdot$ is often referred to as **multiplication,** and we say that we are writing the group multiplicatively. It is extremely important, however, to realize that this "multiplication" is constrained only by the three axioms of the definition and may be very bizarre in comparison to the usual types of multiplication with which we are familiar.

A group may be nicer than usual by virtue of having one or more properties in addition to those legislated in the original definition. The only groups of interest in this text are all **Abelian;** i.e., they all satisfy the following additional constraint:

---

4. $b \cdot c = c \cdot b$ for all choices of b and c from G.

All groups comprehended by Examples II.1 and II.2 are necessarily Abelian; that property is (as was associativity) inherited from the fact that both addition and multiplication of integers are Abelian operations.

The simplest possible type of group is a **cyclic** group: a group $\{G; \cdot\}$ is said to be cyclic if there exists an element g in G such that every element of G is some power of this distinguished element g—which is referred to as a **generator** of the group G. Consider the following two illustrative examples.

Example II.3: $G = \{0, 1, 2, 3, 4, 5\}$ and $\cdot$ = addition modulo 6. I claim that the element 1 is a generator for this (therefore) cyclic group. To say that an element h of G is a power of g—using multiplicative notation—is to say that $h = g^n = g \cdot g \cdot g \cdot \ldots \cdot g$ for some nonnegative[1] integral number of factors n. In additive notation, this requirement takes the notational form h = "ng"[2] = $g + g + g + \ldots + g$. In our current example $1 = 1, 2 = 1 + 1, 3 = 1 + 1 + 1, 4 = 1 + 1 + 1 + 1, 5 = 1 + 1 + 1 + 1 + 1$ and $0 = 1 + 1 + 1 + 1 + 1 + 1$, which proves that G is a cyclic group with the element 1 as a generator. Notationally we indicate this by writing $G = <1>$.

Our next example is much more interesting than Example II.3, in spite of the fact that it is Example II.3 all over again—in a sense to be made precise shortly.

Example II.4: Take $G = \{1, 2, 3, 4, 5, 6\}$, and take the binary operation to be multiplication modulo 7. Then it is straightforward to verify that, for instance, the element 5 can serve as a generator of this Abelian group. For instance, $5^6 = 15{,}625 \equiv 1 \pmod 7$, so that $1 = 5^6$ in our current group. This information has the same mathematical content as does the information contained in the statement that $\{5, 25, 125, 625, 3{,}125, 15{,}625\}$ constitutes a reduced residue system modulo 7 (see page 50, problem 1(b)). Another way of interpreting the fact that this particular group is cyclic is to realize that it corresponds exactly to the fact that 5 is a primitive root of 7.

It is time to explain why Example II.3 and Example II.4 are the same example, from the point of view of abstract algebra. We need the following basic definition.

Definition II.2: Two groups $\{G; \cdot\}$ and $\{G'; \cdot'\}$ are said to be **isomorphic** if and only if there exists a one-to-one correspondence[3] f, say, between the two sets G and G' satisfying the condition $f(b \cdot c) = f(b) \cdot' f(c)$ for every choice of b and c in G.

Remarks:

1. The word isomorphic means, literally, "same form," so to say that two groups are isomorphic is to say that they look the same to an algebraist.

[1]Negative integral values of n also make sense, in terms of inverses.

[2]WARNING: The term "ng" *doesn't* mean multiplication of two elements of G.

[3]Another term for this is bijection.

2. Two groups are isomorphic if they can be made to look exactly the same by the changing of labels. We illustrate this in the following example.

Example II.5 (Example II.3 = Example II.4): To prove that the group of Example II.3 is isomorphic to the group of Example II.4, it is enough to define the function $f{:}G \to G'$ by $f(1) = 5$ and apply the condition given in Definition II.2 repeatedly to learn that $f(2) = f(1 + 1) = f(1) \cdot' f(1) = 5 \cdot' 5 = 25 = 4$, $f(3) = 6$, $f(4) = 2$, $f(5) = 5$, and $f(6) = f(0) = 1$. Each of the two groups is isomorphic to the cyclic group of order 6, and f provides an explicit **isomorphism** between these two groups.

To convince you beyond a reasonable doubt that elementary number theory and elementary abstract algebra are "isomorphic," it is necessary first to do a little elementary abstract algebra. So far, we have defined group, introduced the special classes of Abelian and cyclic groups, and defined isomorphism. Let's now go after a couple of other things—subgroups, and the idea of the order of a group element—and then immediately relate these important algebraic concepts to number theory.

Definition II.3: We say that $\{H; \cdot\}$ is a **subgroup** of the group $\{G; \cdot\}$ provided that H is a subset of G and $\{H; \cdot\}$ is a group.

Definition II.4: An element h of a group $\{G; \cdot\}$ is said to have **order** n if the cyclic subgroup $H = \langle h \rangle$ generated by the single element h contains n elements.

Remarks:

1. It is extremely important to realize that the binary operation associated with the set H is to be the same[1] as that defined on the original group G.
2. I am here allowing myself the liberty of becoming sloppy: a group is (by Definition II.1) an ordered pair, and hence it consists of two parts—the set part and the function (binary operation) part—but we often settle for saying "the group G" for short.
3. The cyclic subgroup $H = \langle h \rangle$ referred to in Definition II.4 consists of all integral powers of the element h—i.e., $H = \langle h \rangle = \{h^0 = e, h^1 = h, h^{-1}, h^2, h^{-2} = (h^{-1})^2, \ldots\}$, where e is the identity element of H.

We have accumulated (barely) enough jargon to state "the first theorem of group theory," namely: Lagrange's Theorem.

[1]Strictly speaking, it is the restriction of the function $\cdot$ to $H \times H$.

Lagrange's Theorem: If G is a finite group and H is a subgroup of G, then the number of elements in the set H is a divisor of the number of elements in the set G.

Sketch of Proof: Let m be the number of elements of G, and let n be the number of elements of H. Then we need to establish that $n|m$. The idea of the proof is to use H to partition G. More precisely, it isn't too hard to show that the set $gH = \{gh_1, gh_2, \ldots, gh_n\}$ consists of precisely n distinct elements of G, regardless of the choice of g in G, and furthermore that g_1H and g_2H are either the same n-element subset of G or are disjoint as sets. Thus G can be realized as a disjoint union of, say, r subsets of size n, and hence $m = nr$.

We are now in a position to show an extremely strong connection between elementary group theory and elementary number theory.

Example II.6: Take as our finite group $\{R; \cdot\}$ of Example II.2 on page 112. We know (pages 48–49) that the number of elements in our group is $\phi(m)$. Let r be a generic element of R, and take H to be the cyclic subgroup of $G = R$ generated by r; i.e., $H = <r>$. Lagrange's Theorem tells us that the order—call it n—of the element r (equivalently, the number of elements in the subgroup $H = <r>$) must be a divisor of $\phi(m)$. So $r^n = 1$ in the group R, which allows us to conclude that $r^n \equiv 1 \pmod{m}$ and finally that $r^{\phi(m)} \equiv 1 \pmod{m}$—the Euler-Fermat Theorem of Section 3.2 (page 49)! We have learned that the extremely useful Euler-Fermat Theorem of elementary number theory can be obtained as a corollary of Lagrange's Theorem in elementary group theory. The definition of $\mathrm{ord}_m b$—the order of b modulo m (Section 3.5, page 66)—is entirely consistent with the Definition II.4, the order of an element in an abstract group.

If we were to do a little more group theory, a natural next step would be to define what is known as the **quotient (factor) group** of the group G by the subgroup H of G, denoted G/H. It would then be possible, for instance, to realize that every complete residue system modulo m can be regarded as a set of **coset representatives** for the quotient group Z/mZ. Rather than follow this course, I will devote the rest of this appendix to introducing the notions of *ring, the group of units of a ring,* and *field.*

Definition II.5: A **ring** is an ordered triple $\{R; +, \cdot\}$ consisting of a set of elements R together with two functions $+$ and $\cdot$ from $R \times R$ into R, subject to the following three axioms:

1. $\{R; +\}$ is an Abelian group.
2. $b \cdot (c \cdot d) = (b \cdot c) \cdot d$ for all choices of the elements b, c and d from the set R (the binary operation $\cdot$ is **associative**).

3. $b \cdot (c + d) = (b \cdot c) + (b \cdot d)$, and $(b + c) \cdot d = b \cdot d + c \cdot d$, for all choices of the elements b, c, and d from the set R (multiplication is **distributive,** both left and right, with respect to addition).

As already indicated, the first of the two functions (binary operations) is called addition, while the second is called multiplication. All our rings have two further properties:

4. $b \cdot c = c \cdot b$ for all choices of the elements b and c from the set R (**commutativity**).
5. There exists one and only one element, denoted by 1, that satisfies the condition $r \cdot 1 = r$ for all choices of the element r of R. (This unique element 1 is referred to as the **multiplicative identity** of the ring R.)

Example II.7: Let $m > 1$ be our modulus, and let $C = \{0, 1, 2, \ldots, (m - 1)\}$. If we take $+$ to be addition modulo m and $\cdot$ to be multiplication modulo m, then $\{C; +, \cdot\}$ is a commutative ring with multiplicative identity 1. The identity element for the associated additive Abelian group $\{C; +\}$ is the number 0.

Example II.8: We specialize the immediately preceding example by assuming that $m = p =$ a prime. In this case, there is another group lurking around, namely, $\{C - \{0\}; \cdot\}$. This is the prototypical example of a field in elementary number theory. Let's consider the definition in full generality.

Definition II.6: A **field** is an ordered triple $\{F; +, \cdot\}$ satisfying the following four axioms:

1. F contains at least two elements, denoted 0 and 1.
2. $\{F; +\}$ is an Abelian group.
3. $\{F^*; \cdot\}$ is an Abelian group. Here $F^* = F - \{0\}$.
4. Multiplication $(\cdot)$ is distributive over addition $(+)$.

It is not too difficult to show that the ring $\{C; +, \cdot\}$ of Example II.7 is a field if and only if the modulus m is a prime (see page 119, problem 13). In the case of a composite modulus, there is still a fairly satisfactory consolation prize available—but first we need one final concept.

Definition II.7: An element u in a commutative ring R with multiplicative identity 1 is said to be a **unit** of that ring, provided that there exists an element v of R such that $u \cdot v = 1$.

Example II.9: The set of units in the familiar ring $\{Z;\ +,\ \cdot\}$ consists of the two elements 1 and -1 (proof?). Please notice that $\{\pm 1;\ \cdot\}$ is a multiplicative group. That this is not merely an accident is shown by our final theoretical result in this appendix.

Theorem: Assume that $\{R;\ +,\ \cdot\}$ is a commutative ring with multiplicative identity 1. Then the set of units of R constitutes a group with respect to the second binary operation.

Proof: See page 119, problem 14.

Example II.10: The group of units of the ring $\{C;\ +,\ \cdot\}$ is isomorphic to the group $\{R;\ \cdot\}$, where R is any reduced residue system modulo m. In particular, the number of units in the given ring is $\phi(m)$. The structure of this particular unit group—as a finite Abelian (but not, in general, cyclic) group—is very important but, as the saying goes, lies outside the scope of this appendix.

I hope that the above rudimentary introduction to the vast world of abstract algebra has convinced you that there are many extremely strong interconnections between number theory and algebra—and that these should be pursued for the mutual benefit of both subjects.

PROBLEMS

Solutions (or hints at solutions) to the problems marked with an asterisk are given on page 132.

* **1.** *(a) Let the integer $m > 1$ be the modulus, let C denote any arbitrary but fixed complete residue system modulo m, and let $\cdot$ denote addition modulo m (all of this precisely as in Example II.1, page 111). Prove that $\{C;\ \cdot\}$ is always a cyclic group.

(b) Let the integer $m > 1$ be the modulus, let R denote any arbitrary but fixed reduced residue system modulo m, and let $\cdot$ denote multiplication modulo m (all of this precisely as in Example II.2, page 112). Prove that $\{R;\ \cdot\}$ is always an Abelian group.

***2.** In each of the following, decide whether the given statement is true or false, and then prove that your answer is correct.

- ***(a)** Every cyclic group is Abelian.
- ***(b)** Every Abelian group is cyclic.[1]
- ***(c)** The group of a problem 1(a) is always Abelian.
- ***(d)** The group of problem 1(b) is always cyclic.

***3.** In Example II.4 on page 113, it was indicated (without all the details being filled in) that the number 5 had order 6 in the group G.

- (a) Find the order of each element of G.
- ***(b)** Use what you have learned in problem 3(a) to find all other elements of G, if any, that serve as generators for this cyclic group.

4. Prove that the relation of isomorphism is an equivalence relation—i.e., that it is reflexive, symmetric, and transitive (see page 105, Definition I.1.1).

***5.** True or false? Proof?

- ***(a)** If two (finite) groups are isomorphic, they must possess the same number of elements.
- ***(b)** If two (finite) groups have the same number of elements, they must be isomorphic.
- ***(c)** If a group G has order m and $n \nmid m$, then G has no subgroup H of order n.
- ***(d)†** If a group G has order m and $n|m$, then G must have at least one subgroup H of order n.
- ***(e)** Cyclic groups are "too nice" to be "too big."
- ***(f)** H, the cyclic subgroup generated by h—as in Definition II.3 on page 114—is always infinite. Look at the description of H in remark 3, page 114.
- ***(g)** It doesn't make any sense to say that 4 is less than 5 in the context of Example II.1 on page 111.

***6.** ***(a)** Give a full-fledged proof of Lagrange's Theorem (page 115).

- ***(b)** Prove that Fermat's Little Theorem (see Corollary 1 on page 50) can be deduced from Lagrange's Theorem (page 115).

***7.** ***(a)** Is $\{Z; +\}$ a cyclic group? If so, find all possible generators for Z, and prove that you have done so.

- ***(b)** Is $\{Z; +, \cdot\}$ a ring?
- ***(c)** Is $\{Z; +, \cdot\}$ a field? Prove that your answer is correct.

***8.** ***(a)** Is every field a ring? Proof?

- ***(b)** Is every ring a field?[2] Proof?

[1] If both 2(a) and 2(b) were true, we wouldn't need two terms.

[2] As in the case of problem 2, parts (a) and (b), if the answers to both parts of problem 8 were yes, there would be no need for both of the terms ring and field.

*9. Determine the unit group of $S = Z[\sqrt{-10}]$. See page 26, Example 2.2.2.

10. How many generators does the group $\{R; \cdot\}$ have, where $R = \{0, 1, \ldots, (p - 1)\}$, p is a prime, and $\cdot$ denotes multiplication modulo p?

*11. If F is the set of positive real numbers, $+$ is (ordinary) multiplication, and $\cdot$ is defined via $b \cdot c = b^{\log_q c}$, where $q \neq 1$ is any arbitrary but fixed positive real number, then does $\{F; +, \cdot\}$ form a field? Prove that your answer is correct.

12. Prove that if the group $\{R; \cdot\}$ of Example II.10, page 117 is cyclic, then there are $\phi(\phi(m))$ generators, and their product is congruent to 1 modulo m if $m > 6$.

13. Prove that the ring $\{C; +, \cdot\}$ is a field if and only if the modulus m is a prime.

14. Prove that the set of units in a commutative ring with multiplicative identity forms a group with respect to the ring multiplication.

Solutions/Hints to Selected Problems

SECTION 1.1

2. (c) $3{,}587 = (1{,}819 \cdot 1) + 1{,}768$
$1{,}819 = (1{,}768 \cdot 1) + 51$
$1{,}768 = (51 \cdot 34) + 34$
$51 = (34 \cdot 1) + 17$
$34 = (17 \cdot 2) + 0$

3. (c) $17 = 51 - 34 = 51 - (1{,}768 - (51 \cdot 34)) = (35 \cdot 51) - 1{,}768$
$= 35(1{,}819 - 1{,}768) - 1{,}768 = (35 \cdot 1{,}819) - (36 \cdot 1{,}768)$.

5. (a) True. **(b)** False. **(c)** True. **(d)** False. **(e)** True. **(f)** True.

6. (c) Since b is odd, $b = 2k + 1$ for a suitable integer k (see problem 6(b)); so $b^2 - 1 = (2k + 1)^2 - 1 = 4k(k + 1)$. Now invoke problem 6(a) to finish the argument.

7. (b) Make two applications of part 3 of Theorem 1.1.1 (page 2).

8. See LeVeque's *Fundamentals of Number Theory,* page 14, for a complete solution.

10. (a) All nonzero integers b.
(b) $b = 1, c = -1$ or $b = -1, c = 1$. **(c)** No.

13. (a) Yes. Since $m|n$, we have $n = km$ for some positive integer k. Hence, $b^n - 1 = b^{km} - 1 = (b^m)^k - 1^k = (b^m - 1)\{(b^m)^{k-1} + (b^m)^{k-2} + \ldots + 1\}$.

SECTION 1.2

1. (c) $[3{,}587, 1{,}819] = (3{,}587 \cdot 1{,}819)/(3{,}587, 1{,}819)$, according to Theorem 1.2.2 on page 11, and $(3{,}587, 1{,}819)$, from problem 2(c) on page 5.

2. **(a)** There are no integral solutions because the left-hand side of the proposed equation is always even, whereas 17 is odd.
   (b) $x = 6$ and $y = -1$ will work.

4. **(b)** No. If $b = 2$, $c = 3$, and $d = 4$, then $(b, c, d)[b, c, d] \neq bcd$; look back at problems 3(a) and 4(a), for instance.

Problems 5 through 8 form a package.

8. $(x, y) = G$ if and only if $x = GX$ and $y = GY$ for integers X and Y that are relatively prime—i.e., $(X, Y) = 1$. The requirement $x + y = n$ can then be rewritten as $G(X + Y) = n$, which certainly requires $G|n$. If $G|n$, then $X = 1$ and $Y = (n/G) - 1$ will serve to satisfy both of the equations $G(X + Y) = n$ and $(X, Y) = 1$. As a matter of fact, there will be infinitely many pairs of integers x and y that satisfy both the conditions $(x, y) = G$ and $x + y = n$, whenever $G|n$; how would you prove that?

Problems 9 and 10 are very similar in spirit to 8.

13. Property 5: there exist integers x, y, r, and s such that $bx + dy = 1$ and $cr + ds = 1$. Hence, $(bx + dy)(cr + ds) = 1$.

15. $1 + b + b^2 = (b^3 - 1)/(b - 1)$, so $G|(b^3 - 1)$, and so $G|(b^3 + b^c)$. $G\nmid b$ unless $G = 1$ (why?); so, if $c > 3$, then $G|(1 + b^{c-3})$. Iterate until done.

16. **(a)** Yes. **(b)** No. **(c)**† $(F_m, F_n) = F_{(m,n)}$. Parts (a) and (b) can easily be deduced from part (c).

SECTION 2.1

1. **(b)** The positive integer n has exactly one nontrivial divisor if and only if $n = p^2$ for some prime p.

2. **(b)** Suppose n is composite. Then there exist integers b and c such that $n = bc$ with $1 < b \leqslant c < n$. By the Fundamental Theorem of Arithmetic, there exists at least one prime—say, p—such that $p|b$. Hence, $p \leqslant b$. However, $n = bc \geqslant b^2$ tells us that $b \leqslant \sqrt{n}$, so $p \leqslant \sqrt{n}$.

3. **(a)** $b = 288 = 2^5 \cdot 3^2$ and $c = 51 = 3 \cdot 17$, so $(b, c) = 3$ and $[b, c] = 2^5 \cdot 3^2 \cdot 17$.

   Problems **4**, **5**, and **6** are closely interrelated. The G_n sequence of problem 4 is far easier to investigate than is the corresponding f_n sequence of problem 5. (See note 1 on page 34.) Some of the results of problems 4 and 5 can be deduced from problem 6.

6. **(a)** If we write $m = n + r$ for suitable $r > 0$, then $\left(b^{2^n}\right)^{2^r} - 1^{2^r} = \left(b^{2^{m-1}}\right)^2 - 1^2 = \left(b^{2^{m-1}} + 1\right)\left(b^{2^{m-1}} - 1\right)$. If $m = n + 1$, we are done. Otherwise, iterate the process.
   (b) Use part (a) of this problem in conjunction with the fact that $b^{2^m} - 1$ and $b^{2^m} + 1$ are 2 apart.

8. (a) p is a prime if and only if $\tau(p) = 2$.
   (b) $\tau(n)$ is odd if and only if n is a perfect square.
   (c) $\{1, p, p^2, \ldots, p^e\}$ is the set of all positive integral divisors of the pure prime power p^e, and so $\tau(p^e) = e + 1$.

11. Parts (a), (c), and (e) are extremely close in spirit. Parts (b) and (d) are converses of each other.
   (a) False.
   (b) True.
   (c) True: $(b^2 - d^2)$ is a particularly simple integral linear combination of $(b^2 - c^2)$ and $(c^2 - d^2)$.
   (d) False: the minimum counterexample is $p = 11$.
   (e) True: subtraction is an integral linear combination.
   (f) True.

14. Mathematical induction (see Appendix I, page 106) is one approach.

   Another is to look at the binomial coefficient $\binom{n}{k}$.

15. $b = 2$ and $c = 4$. Use canonical decomposition.

16. (a) $\sigma(p^e) = (p^{e+1} - 1)/(p - 1)$.
   (c) Yes.
   (i) The (perfect) squares.

18. Write $b^p = \{(b + c) - c\}^p$ and use the Binomial Theorem (page 108).

SECTION 2.2

1. (a) No: take, for instance, $p = 30$, $b = 6$, and $c = 10$.

3. (a) No: p needn't even be an element of E—let alone an indecomposable element of E!
   (b) No: $p = 10$ is a counterexample.
   (c) No: $2 + \sqrt{-10}$ is a counterexample.
   (d) No: $11 = (1 + \sqrt{-10})(1 - \sqrt{-10})$.

7. Take complex conjugates.

8. $(bc - 10ce)^2 + 10(be + cd)^2 = (b^2 + 10c^2)(d^2 + 10e^2)$.

11. (b) $\alpha = \beta = (1 + \sqrt{5})/2$. (c)† Yes.

SECTION 2.3

1. Use canonical decomposition.

2. (b) $p \nmid 360$, so $360 = 360 \cdot p^0$, and thus $|360|_p = p^{-0} = 1$.

3. (a) 25, 1/5, 1/125, 125. (b) $1/2^n$. (c) 1.

4. (a) No. (b) Yes. (c) No.

5. **(a)** $|p^k|_p = p^{-k} = 1/p^k$ tends to zero (relative to the usual absolute value) as k tends to plus infinity.
(c) Allow yourself to make use of problem 5(d)†.

8. **(a)** No. **(b)** Yes.

9. **(a)** $+\infty$. **(b)** 0. **(c)** Doesn't exist.

SECTION 3.1

1. **(e)** $9b \equiv 12 \pmod{15}$ if and only if $3b \equiv 4 \pmod 5$ because $(3, 5) = 1$. $b = 3$ works, as do all integers $b \equiv 3 \pmod 5$, subject to the constraint $|b| < 50$. Could we have missed any solutions this way?
(f) No solutions because $(9, 15) = 3 \nmid 13$. $9b \equiv 13 \pmod{15}$ if and only if there exists an integer c such that $9b + 15c = 13$ (why?); and this is impossible because the left-hand side is divisible by 3, whereas the right-hand side is not.

3. **(c)** Regardless of the integer n, it is possible to express n in the form $n = 3k + r$ for suitable integers k and r, with r coming from the set $\{-1, 0, 1\}$ (why?). Hence, $n^3 = (3k + r)^3 = 9K + r^3$ for an appropriate integer K and $r^3 = -1, 0$, or $+1$, and thus $r^3 \equiv 8, 0$, or $1 \pmod 9$ as desired.

5. **(b)** Yes.

6. **(c)** $x \equiv 7 \pmod{10}$, $x \equiv -4 \pmod{-5}$ or, for instance, $x \equiv 1 \pmod 5$.

7. **(a)** True. **(b)** False. **(c)** True. **(d)** True. **(e)** True. **(f)** True. **(g)** True. **(h)** True.

8. **(b)** $X \equiv 10 \pmod{12}$.

9. Every integer x satisfies $x \equiv b \pmod{12}$ for precisely one value of b coming from the set $\{0, 1, 2, 3, 4, 5, 6, 7, 8, 9, 10, 11\}$. If $b = 0, 2, 4, 6, 8$, or 10, then $x \equiv 0 \pmod 2$. If $b = 3, 6$, or 9, then $x \equiv 0 \pmod 3$. If $b = 1$ or 5, then $x \equiv 1 \pmod 4$. If $b = 11$, then $x \equiv 5 \pmod 6$. Finally, if $b = 7$, then $x \equiv 7 \pmod{12}$. The integer 6 satisfies the first two of these congruences.

10. In the language of abstract algebra, the mapping $\sigma: Z[X] \to Z$ given by $\sigma(f[X]) = f(b)$, for some fixed integer b, is referred to as the **evaluation** (or **substitution**) **homomorphism.**

12. Assume $m > 4$ and m composite. Then it must be possible to write $m = bc$ for suitable integers b and c satisfying $1 < b \leqslant c < m$. Then, if $b < c$, we have $(m - 1)! = 1 \cdot 2 \cdot \ldots \cdot b \cdot \ldots \cdot c \cdot \ldots \cdot (m - 1) \equiv 0 \pmod m$ because $m = bc|(m - 1)!$. What about the case $b = c$?

13. **a.** Yes. **(b)** $m = 0$.

14. **(a)** $b \equiv b \pmod m$ if and only if $m|(b - b)$, if and only if $(b - b)/m$ is an integer, if and only if $|(b - b)/m|_p \leqslant 1$ for all primes p, if and only if $|0|_p \leqslant 1$ for all primes p; and $|0|_p = 0 \leqslant 1$ for all primes p. The congruence relation is symmetric because $|(b - c)/m|_p \leqslant 1$ for all primes p implies that $|(c - b)/m|_p \leqslant 1$ for all primes p. Now, $|(b - d)/m|_p = |(b - c)/m + (c - d)/m|_p \leqslant \max\{|(b - c)/m|_p, |(c - d)/m|_p\}$, by the Ultrametric Inequality, and so the transitive property of the congruence relation holds. See also problem 17.

Problems **15** and **16** can be regarded as variations on a theme—the theme being Wilson's Theorem.

18. See also problem 14(a) and problem 17. A rational number q is an integer if and only if the canonical decomposition of q is of the form $q = p_1^{e_1} \cdot p_2^{e_2} \cdot \ldots \cdot p_k^{e_k}$, where $e_i > 0$ for all values of i with $1 \leqslant i \leqslant k$. Under such circumstances, $|q|_p = 1$ for all primes p that fail to enter the canonical decomposition of q and $|q|_p < 1$ for all values of i with $1 \leqslant i \leqslant k$.

SECTION 3.2

1. (b) $5 \equiv 5 \pmod 7$, $25 = 5^2 \equiv 4 \pmod 7$, $125 = 5^3 = 5 \cdot 5^2 \equiv 5 \cdot 4 \equiv 6 \pmod 7$, $625 = 5^4 = 5^2 \cdot 5^2 \equiv 4^2 \equiv 2 \pmod 7$, $3{,}125 = 5^5 = 5^4 \cdot 5 \equiv 2 \cdot 5 \pmod 7$, and $15{,}625 = 5^6 = (5^3)^2 \equiv 6^2 \equiv 1 \pmod 7$, so these six consecutive positive integral powers of 5 are pairwise incongruent modulo 7 and, modulo 7, behave precisely like the set $\{1, 2, 3, 4, 5, 6\}$ and hence constitute a reduced residue system modulo 7.

2. (b) $\{3, 3^2, 3^3, 3^4, 3^5, 3^6\}$. Compare problem 1(b).

3. (b) No, because of the exceptional case of $m = 1$; but yes, for all $m > 1$, because $(1, m) = (m, m - 1) = 1$.

(c) $(r, m) = 1$ implies (and is implied by) $(m - r, m) = 1$, for suppose that $(m - r, m) = G > 1$. Then $G|(m - r)$ and $G|m$, so that $G|\{(-1)(m - r) + (+1)(m)\} = r$; and now $1 < G|(r, m) = 1$. Thus the elements of R can be paired off, and so $^{\#}(R) = \phi(m)$ is even, provided that this pairing is nondegenerate. The answer to Hint 2 is no, which tells us that no pair of elements of R reduces to a single element of R.

(d) $r + (m - r) = m$, and this happens $\phi(m)/2$ times.

5. (a) No. **(b)** No.

6. (d) $\{1, 3, 5, 7\}$.

7. (a) No. **(b)** No. **(c)** No.

9. (c)[†] This problem can be done most elegantly by using Lagrange's Theorem from elementary group theory (see Appendix II, page 115). The idea is to exhibit a subgroup of order 4 within the group $\{Z/\phi(F_k)Z; \cdot\}$.

10. **(c)** From problem 10(a), we may assume (without loss of generality) that $y = 2Y$, and so we have $x^2 + 4Y^2 = z^2$. Both x and z must be odd (why?), and so $x^2 \equiv z^2 \equiv 1 \pmod 8$. Thus $1 + 4Y^2 \equiv 1 \pmod 8$, making $Y^2 \equiv 0 \pmod 2$. It follows that $Y \equiv 0 \pmod 2$.

(d) Yes.

(e) No.

(f) No.

12. **(a)** No. **(b)** Yes. **(c)** No.

13. **(a)** True. **(b)** True. **(c)** True. **(d)** True. **(e)** True. **(f)** True.

SECTION 3.3

1. **(b)** $G = (b, m) = (15, 30) = 15 \nmid 10 = c$ so the given linear congruence isn't solvable.

(d) $(64, 105) = 1 | 83$ so this congruence is solvable.

2. **(c)** Precisely those integers x satisfying $x \equiv -43 \pmod{105}$.

(d) Precisely those integers x satisfying $x \equiv -1$ or 6 or $13 \pmod{21}$.

Please notice that, if a linear congruence is solvable, there are always infinitely many integers that satisfy the given linear congruence. In spite of that, Definition 3.3.1 (page 55) has been arranged very carefully, so that the number of solutions of any linear congruence is always finite. See, for instance, the answers to the following problems 3 and 4.

3. **(a)** Three. **(e)** Six.

4. **(a)** $\{-1, 6, 13\}$; compare the answer to problem 2(d).

(e) $\{-96, -61, -26, 9, 44, 79\}$.

6. **(a)** 2, 3.

(d) There are no solutions to this quadratic congruence. See page 58, problem 9, for more information.

7. **(a)** Two. **(b)** Four. **(c)** Zero. **(d)** Zero. **(e)** Zero. **(f)** Zero.

8. **(d)** $(8X + 1)^2 \equiv 225 \equiv 15^2 \pmod{13}$ is solvable.

9. One of the many recurring themes in elementary number theory is pairing processes: $(p - 1)! = \prod_{r=1}^{\frac{p-1}{2}} r(p - r) \equiv (-1)^{(p-1)/2} \cdot \prod_{r=1}^{\frac{p-1}{2}} r^2 \equiv -1 \pmod p$. If $p \equiv 1 \pmod 4$, then $\left\{\prod_{r=1}^{\frac{p-1}{2}} r\right\}^2 \equiv -1 \pmod p$. If $p \equiv 3 \pmod 4$ and $z^2 \equiv -1 \pmod p$, then raise this congruence to the power $(p - 1)/2$ and get a contradiction upon applying Fermat's Little Theorem.

SECTION 3.4

1. **(c)** $(5, 10) \nmid -2$ so there are no solutions.
   (d) $5X \equiv 6 \pmod 7$ is equivalent to $X \equiv 4 \pmod 7$ (why?), and $4X \equiv 5 \pmod 9$ is equivalent to $X \equiv 8 \pmod 9$. $x = 53$ is the only solution of both congruences that also satisfies $0 < x \leqslant 100$.
2. If $X \equiv b \pmod m$ is solvable, then $X \equiv b \pmod{p^e}$ is solvable for every p-component p^e of m because $p^e | m$. The converse is true because of the Chinese Remainder Theorem: the p-components of m are pairwise relatively prime, and we can apply Corollary 3.4.1, given on page 61.
3. $19X \equiv 1 \pmod 4$ is equivalent to $X \equiv -1 \pmod 4$, $19X \equiv 1 \pmod 5$ is equivalent to $X \equiv -1 \pmod 5$ and $19X \equiv 1 \pmod 7$ is equivalent to $X \equiv -4 \pmod 7$ so we may take $b_1 = -1$, $b_2 = -1$, and $b_3 = -4$ in Corollary 3.4.1 (page 61). We may also take $c_1 = -1$, $c_2 = +2$, and $c_3 = -1$, and thus $x_0 = 59$.
4. $-2 + 3^4 = 79$, $-4 + 5^6 = 15{,}621$. See also Paul Halmos' provocative article "Does Mathematics Have Elements?" cited in the Bibliography.
6. $X \equiv 1 \pmod 7$, $X \equiv 1 \pmod 5$, and $X \equiv 0 \pmod 4$.
11. **(a)** Yes. **(b)** No.

SECTION 3.5

1. **(c)** $\phi(km)$. **(d)** $k\phi(m)$. **(e)** $\phi(km) \leqslant k\phi(m)$ for all positive integers k and m. The constraint of (d) isn't as strong as—strictly speaking, is no stronger than—the constraint of (c).
2. **(a)** $p_1^{e_1-1} \cdot p_2^{e_2-1} \cdot \ldots \cdot p_k^{e_k-1} \cdot (p_1 - 1) \cdot (p_2 - 1) \cdot \ldots \cdot (p_k - 1) = 2^2$ requires $k \leqslant 3$ and $e_i \leqslant 3$ for all i with $1 \leqslant i \leqslant k \leqslant 3$. $2^{\alpha-1} \cdot p^{\beta-1} \cdot q^{\gamma-1} \cdot (p - 1) \cdot (q - 1) = 2^2$. If the prime 5 appears in the canonical decomposition of x, then $x = 5$ or 10. If $3 | x$, then $x = 12$. The only other value of x is $x = 2^3 = 8$.
   (b) All solutions of $\phi(x) = 20 = 2^2 \cdot 5$ can systematically be found by applying the formula for ϕ given in Theorem 3.5.2 (page 65). The main thing is to be sure to consider all the logical possibilities.
3. **(a)** False. **(b)** True. **(c)** False. **(d)** True. **(e)** True. **(f)** True. **(g)** True.
4. **(c)** $3 \nmid 1{,}093$ and $1{,}093$ is a prime (why?). $\text{ord}_{1,093} 3 = 7$.
6. $\phi(m) = \text{ord}_m(b \cdot q) + r$, with $0 \leqslant r < \text{ord}_m b$. $1 \equiv b^{\phi(m)} = (b^{\text{ord}_m b})^q b^r \equiv 1^q \cdot b^r \equiv b^r \pmod m$ implies $r = 0$.
7. **(a)** Yes: $12 \equiv 5 \pmod 7$.
   (b) No.

(c) h is a primitive root modulo m.
(d) Yes: (d) implies (c) implies (e).
(e) No.

9. (a) Yes: use the explicit formula for ϕ given in Theorem 3.5.2 (page 65).
   (b) No: put $d = 6$ and $n = 4$, for instance.
10. When x gets large, at least one p-component of x must get big, and hence $\phi(x)$ must get big.
11. Use canonical decomposition and Theorem 3.5.2 (page 65).
13. (a) Yes. (b) No.
14. Problem 2 is **much** easier than Problem 1 and can be solved via Theorem 3.5.2 (page 65). See also page 79, note 8.
15. $1 + b + b^2 = (b^3 - 1)/(b - 1)$ and $\text{ord}_p(1 + b) \neq 1, 2,$ or 3.
16. $(b^u)^2 - 1 \equiv 0 \pmod{p^e}$. Factor the difference of two squares.
18. $\text{ord}_m b \mid \phi(m)$, and yet $\text{ord}_m b = m - 1$.
19. (b) Write $[h, k] = hH = kK$ so that $(H, K) = 1$. $(b^h)^H \cdot (c^k)^K \equiv 1 \pmod{m}$.
20. Yes.
27. Yes.
30. Mathematical induction is one way to go; see Appendix I, page 106.
31. No.

SECTION 3.6

1. (b) $b \equiv c \pmod{m}$ implies $\text{ord}_m b = \text{ord}_m c$.
   (e) $g' = g + x$ for some integer x, so $g' \equiv g \pmod{p}$, and g was a primitive root modulo p.
   (g) This result can be established by mathematical induction (see page 106 in Appendix I), starting from the fact that $b^2 \equiv 1 \pmod{2^3}$, regardless of the odd integer b.
2. Yes: $\psi(d) = \phi(d)$ for all $d \mid (p - 1)$. See also page 69, problem 4.
4. (a) Yes: apply property 4 of the order symbol (page 67).
   (b) Use part (a) of this problem. For $p = 7$, the primitive roots are 3 and 5. Thus there are $2 = \phi(\phi(7))$ of them.
5. Yes. Compare page 103, problem 7.

SECTION 3.7

1. For $m = 13$, $1^2 \equiv 1 \pmod{13}$, $2^2 \equiv 4 \pmod{13}$, $3^2 \equiv 9 \pmod{13}$, $4^2 \equiv 3 \pmod{13}$, $5^2 \equiv 12 \pmod{13}$, and $6^2 \equiv 10 \pmod{13}$; so the elements of R are 1, 3, 4, 9, 10, and 12. For $m = 6$, the only element of R is the number 1.

2. $b = 1$ and $b = 8$.

5. $n = 2$ and $m = p$, so $\phi(m) = \phi(p) = p - 1$, $(n, \phi(m)) = (2, p - 1) = 2$, and thus $\phi(m)/(n, \phi(m)) = (p - 1)/2$.

6. **(a)** No. **(b)** No.

7. $b^{p-1} \equiv 1 \pmod p$ so $p|\{(b^{(p-1)/2})^2 - 1^2\}$. Now factor the difference of squares.

8. **(a)** ind 2 = 1, ind 4 = 2, ind 8 = 3, ind 7 = 4, ind 5 = 5, and ind 1 = 6.

10. The "obvious" way to do the evaluation is to perform the addition inside the "absolute value" bars first. Page 32, problem 4(d) provides an alternate method of attack.

SECTION 4.1

1 **(a)** 6, 6, 6, 6. **(c)** 91, 4, 4, 4.

2 **(a)** Lagrange's Theorem is applicable for each of the three moduli because $f(X)$ has leading coefficient 1. The upper bound is (therefore) uniformly 6.

(b) If $p = 3$ or 7, then the upper bound is 4. If $p = 2$, Lagrange isn't applicable. The degree of the polynomial congruence $f(X) = 2X^4 + 6X^3 - 7 \equiv 0 \pmod 2$ is zero, as is the number of solutions of it.

3. (c) $\{0, 1\}$ for $p = 2$, again $\{0, 1\}$ for $p = 3$, and $\{0, 1, 2, 4\}$ for $p = 7$.

4 **(a)** No: recall $X^2 \equiv 1 \pmod 8$.

(b) Yes: a combination of Lagrange's Theorem and Definition 4.1.1 (page 82).

5 **(a)** Yes. **(b)** No.

8. (b) $-X^2 - X$. **(c)** $3X^2 + 3X + 1$.

9. (a) No: take $p = 3$ and $b = 5$, for instance.

(b) No: Fermat's Little Theorem, in slightly modified form, shows that $|b^p - b|_p \leq 1/p$ for <u>all</u> integers b and for all primes p.

(c) Yes: compare part (b) of this problem; furthermore, it has been assumed <u>twice</u> that b is an integer.

SECTION 4.2

1. **(a)** Neither $3t \equiv -1 \pmod 3$ nor $15t \equiv -7 \pmod 3$ is solvable.

(b) $9t \equiv -3 \pmod 3$ has solutions $t = 0$, 1, and 2.

(c) $f(4) = 3^3$, $f'(4) = 3^2$, $f(13) = 3^3 \cdot 7$, $f'(13) = 3^3$, $f(22) = 3^3 \cdot 19$, and $f'(22) = 3^2 \cdot 5$. The resulting three linear congruences—namely, $3^2t \equiv -1 \pmod 3$, $3^3t \equiv -7 \pmod 3$, and $3^2 \cdot 5t \equiv -19 \pmod 3$—are all insoluble.

3. $x_0 + ip \equiv x_0 \pmod{p}$, $0 \leqslant i \leqslant p - 1$, implies $f(x_0 + ip) \equiv 0 \pmod{p^2}$ for one or more of these i values.

4. **(b)** See page 92, problem 1.

8. $0.b_0 b_1 b_2 \ldots$, where $b_0 = -2$, for instance.

9. See page 24, problem 14. $\binom{i}{k} = i!/k!(i-k)! = i(i-1) \cdot \ldots$ $\cdot (i - k + 1)/k!$

SECTION 4.3

1. See page 91, problem 4(b).

SECTION 5.1

1. **(a)** Proof by contradiction: if $1 \leqslant i < j \leqslant (p-1)/2$ and $j^2 \equiv i^2 \pmod{p}$, then $p|(j^2 - i^2) = (i+j)(j-i)$, which implies, since p is a prime, that $p|(j+i)$ or $p|(j-i)$. However, $0 < j - i < j + i < p$.
   (b) $b = pq + r$ with $1 \leqslant r < p$ (Division Algorithm), and $(b/p) = (r/p)$. In addition, $r^2 \equiv (p-r)^2 \pmod{p}$, so that, if r is too big, it can be replaced by $p - r$.
   (d) In case $p = 13$, the requisite elements are 1, 4, 9, 3, 12, and 10.
   (e) In case $p = 11$, the requisite elements are the integers in the closed interval [1, 10] other than 1, 4, 9, 5, and 3.

2. **(b)** No.

3. **(a)** $S = \{10, 20, 30, 40, 50, 60\}$, $n = 4$, so that 10 is a quadratic residue of 13, in agreement with 1(d), for instance.

5. **(b)** $-n \equiv n \pmod{2}$ and $[2k/p] = 0$ for all k, with $1 \leqslant k \leqslant (p-1)/2$.

6. **(a)** Property 3 on page 95 gives $(-1/p) \equiv (-1)^{(p-1)/2} \pmod{p}$, but each of $(-1/p)$, $(-1)^{(p-1)/2}$ is ± 1, so the only way their difference can be divisible by the <u>odd</u> prime p is for their difference to be zero.
   (b) $1^2 \equiv -1 \pmod{2}$. Let p be an odd prime. Then -1 is a quadratic residue of p if and only if $(-1/p) = (-1)^{(p-1)/2} = +1$ if and only if $(p-1)/2$ is even.
   (c) Yes: there are infinitely many primes $p \equiv 1 \pmod{4}$.

7. **(b)** Yes: compare problem 6(c).

8.† **(a)** Yes. **(b)** Yes.

SECTION 5.2

1. **(b)** $+1, -1$.

2. $X^2 \equiv p \pmod{q}$ having no solutions becomes $(p/q) = -1$ in the language of Chapter 5. "LQR" (the Law of Quadratic Reciprocity) tells us that $(p/q)(q/p) = -1$ whenever $p \equiv q \equiv 3 \pmod{4}$.

3. No: proof by contradiction. Does the result of this problem contradict the fact that the product of two quadratic nonresidues is a quadratic residue? Why?

4. (a) For any $p \neq 5$, $(5/p) = (p/5)$, so the answer is all $p \equiv \pm 1 \pmod 5$.

6. The set $\{1 \cdot p, 2 \cdot p, \ldots, (p-1) \cdot p\}$ is a reduced residue system modulo p. The number of quadratic residues modulo p = the number of quadratic nonresidues modulo p.

7. Compare page 74, problem 5. $3^2 \equiv +1 \pmod{2^{2k}+1}$, and 2^{2k-1} is the least positive integer that will do the job (i.e., will leave a remainder of $+1$ upon division by p) iff $3^{(p-1)/2} \equiv -1 \pmod p$.

10. $g^2 g^4 \cdot \ldots \cdot g^{p-1} = (g^2)^{(1+2+\ldots+(p-1)/2)}$. $1 + 2 + \ldots + (p-1)/2 = (p^2-1)/8$. $(p^2-1)/4 = (p-1)[(p+1)/4]$ is a legitimate integral rewrite precisely half the time.

11. (a) $2^{4p} \equiv 1 \pmod{(4p+1)}$, and you need to establish the minimality of $4p$. $2^4 \equiv 1 \pmod q$. Prove that $2^{2p} \equiv -1 \pmod q$ by applying a combination of Euler's Criterion (page 95) and "the quadratic character of 2"—Corollary 5.1.1 on page 97 and problem 7 on page 98.

12. For parts (a) and (c), the picture can be very helpful—see page 104, note 3. This proof of "LQR" is approximately discrete double/iterated integration. There are no lattice points on the line $y = (p/q)x$ because $(q, p) = 1$ and $0 < x < p$.

13.† I don't know the answer to the question but I believe that the answer is yes. See Bibliography, page 134.

14.† $1 + x + x^2 + x^3 + x^4 = y^2$ requires that $x^2 < y < x^2 + x$. Write y in base x^2.

APPENDIX I

1. (b) $(k+1)(k+2)(k+3)(k+4) = k(k+1)(k+2)(k+3) + 4(k+1)(k+2)(k+3)$.

(g) See the hint for problem 14 on page 24.

(h) $(k+1)! - 1 + (k+1)(k+1)! = (k+1)![1 + (k+1)] - 1$.

3. (b) $b \sim b$ if and only if $b \neq 0$.

(c) $b \sim b$ if and only if $b = 0$.

4. (a) 10, $10!/5!5! = (10 \cdot 9 \cdot 8 \cdot 7 \cdot 6)/(5 \cdot 4 \cdot 3 \cdot 2) = 252$, 91. **(b)** Yes.

(c) Yes. **(e)** $\binom{52}{13}$.

5. "Counting problems" can be very tricky because of various ways of overcounting, undercounting, etc. Here there is a uniformity—each leg is counted with multiplicity three.

APPENDIX II

1. **(a)** The number 1 always serves as generator.
2. **(a)** True. **(b)** False. **(c)** True. **(d)** False.
3. **(b)** 3 is the only other generator of this group. $\phi(\phi(7)) = 2$.
5. **(a)** True.
   (b) False (highly!).
   (c) True.
   (d)† This proposed converse of Lagrange's Theorem is false in general, but it is true, for instance, if we make the additional assumption that G be Abelian.
   (e) True: no cyclic group can be any bigger than Z.
   (f) False.
   (g) True.
6. **(a)** Work from the sketch of proof given on page 115.
   (b) Implication is transitive.
7. **(a)** Yes. $\pm$ 1. **(b)** Yes. **(c)** No.
8. **(a)** Yes. **(b)** No: use Z as a counterexample.
9. ± 1.
11. Yes.

Bibliography

BOOKS

Bachman, G. *Introduction to* p-*adic Numbers and Valuation Theory.* New York: Academic Press, 1964.

Birkhoff, G., and Maclane, S. *A Survey of Modern Algebra*. Rev. ed. New York: Macmillan, 1959.

Borevich, Z. I., and Shafarevich, I. R. *Number Theory.* New York: Academic Press, 1966.

Cohn, H. *A Second Course in Number Theory.* New York: John Wiley, 1962.

Davenport, H. *The Higher Arithmetic*. New York: Dover, 1983. Originally published in 1952.

Hardy, G. H., and Wright, E. M. *An Introduction to the Theory of Numbers*. 5th ed. London: Oxford University Press, 1979.

Herstein, I. *Topics in Algebra*. 2d ed. Lexington, Mass.: Xerox 1975.

LeVeque, W. J. *Fundamentals of Number Theory.* Cambridge, Mass.: Addison-Wesley, 1977.

Niven, I., and Zuckerman, H. S. *An Introduction to the Theory of Numbers*. 4th ed. New York: John Wiley, 1980.

Pollard, H., and Diamond, H. G. *The Theory of Algebraic Numbers*. The Carus Mathematical Monographs, No. 9. MAA: Baltimore, Md., 1950.

MAA Studies in Mathematics, *Studies in Number Theory.* vol. 6, edited by W. J. LeVeque. Englewood Cliffs, N.J.: Prentice-Hall, 1969.

Weyl, H. *Algebraic Theory of Numbers*. Princeton, N.J.: Princeton University Press, 1940.

JOURNAL ARTICLES

Edgar, H. M. "The Exponential Diophantine Equation $1 + a + a^2 + \ldots + a^{x-1} = p^y$." American Mathematical Monthly 81, no. 8 (1974): 758–59.

Filaseta, M. "An Application of Faltings' Results to Fermat's Last Theorem." C. R. Math. Rep. Acad. Sci. Canada VI, no. 1 (1984): 31–32.

Gerstenhaber, M. "The 152-nd Proof of the Law of Quadratic Reciprocity." American Mathematical Monthly 70 (1963): 397–98.

Halmos, P. R. "Does Mathematics Have Elements?" The Mathematical Intelligencer 3, no. 4 (1981): 147–52.

Heath-Brown, D. R. "Artin's Conjecture for Primitive Roots." Quart. J. Math. Oxford (2), 37 (1986): 27–38.

Mazur, B. "Arithmetic on Curves." Bulletin of the American Mathematical Society 14, no. 2 (1986): 207–59.

Ore, O. "On the Averages of the Divisors of a Number." American Mathematical Monthly 55, no. 9 (1948): 615–19.

Sholander, M. "Least Common Multiples and Highest Common Factors." American Mathematical Monthly 68, no. 10 (1961): 984.

Wagon, S. "Fermat's Last Theorem." The Mathematical Intelligencer 8, no. 1 (1986): 59–61.

Index

Boldface numbers indicate a definition or statement of theorem.